DEPRAVED INDIFFERENCE: *the* WORKERS' COMPENSATION SYSTEM

DEPRAVED
INDIFFERENCE:
the WORKERS'
COMPENSATION SYSTEM

Patrice Woeppel, Ed.D.

iUniverse, Inc.
New York Bloomington Shanghai

depraved INDIFFERENCE: *the* **Workers' Compensation System**

iUniverse books may be ordered through booksellers or by contacting:

iUniverse
1663 Liberty Drive
Bloomington, IN 47403
www.iuniverse.com
1-800-Authors (1-800-288-4677)

Because of the dynamic nature of the Internet, any Web addresses or links contained in this book may have changed since publication and may no longer be valid.

Cover design by Jamal Berkeley

ISBN: 978-0-595-48373-0 (pbk)
ISBN: 978-0-595-60464-7 (ebk)

Printed in the United States of America

This book is dedicated to the workers who have lost their health or their lives through their work, and to the families who live with the pain and the grief of this American tragedy.

CONTENTS

Acknowledgments

I am forever grateful to my husband, Don Rose and my son, Jamal Berkeley, whose love and encouragement and belief in my abilities kept me going, while I battled my own fight with workers' compensation.

I am deeply indebted to all the people who allowed me into their lives, or into the life of their beloved family member, to be interviewed for this book. Special thanks to Tammy Miser and to Carol Ann Stoughton: Carol was the first person I interviewed. Each has been an invaluable resource.

Special thanks to my son, Jamal, for creating the cover for this book; to my husband, Don, for his technical assistance; to Sally Axelrod for her professional assistance in editing portions of this book, and for her guidance throughout the process of writing this book. Special thanks also to Beverly Harris, Jefferson Rose and Joseph A. Rose for reading, reviewing and editing portions of the book. Special thanks to Elizabeth Shelton at the Galveston Bay Information Center for her diligence, and knowledge of resources re toxic spills in the Galveston Bay area; to Dr. Marion Moses, and Juliann Sum, J.D., Sc.M, for sharing of their expertise and their work; and to the many, many other people who shared their expertise throughout the writing of this book.

My thanks to National Fire Protection Association for allowing me to reprint with their permission from NFPA 484–2006, *Combustible Metals,* copyright © 2006 NFPA, and NFPA 51B-2003. *Fire Prevention During Welding Cutting and Other Hot Work.* This reprinted material is not the complete official position of the NFPA above referenced subjects, which is represented solely by the standard in its entirety.

I am deeply grateful for the work of Dr. Andrew Charni, my chiropractor, without whom I would never have been well enough to produce this book.

There are so many people to whom I am grateful for their assistance during the process of writing this book. I thank you, even if your name is not mentioned here.

At this point, I must add the usual disclaimer that none of the people, or organizations, whom I have acknowledged are responsible for what I have written.

Finally, I want to thank two very brave and unconventional women, who have inspired many who have come after them: Mamie Till Mobley, mother of Emmett Till, who mobilized her grief into a powerful national fight for justice; and Mary Harris Jones, Mother Jones, the union organizer and orator, who reminds us to "pray for the dead and fight like hell for the living."

Introduction

Every eight minutes in this country, someone dies from an occupational illness or injury. All across America, in every state, across industries and occupations, workers are being injured or killed on the job or exposed to toxic chemicals from which they are dying. In 2004, over 4.3 million workers were reported injured on the job. In the same year, over 5700 workers died from their injuries.[1] An additional estimated 60,000 die each year from occupational illnesses: a national tragedy of pandemic proportions.[2]

For those who remain, whether they will fall into the nether world of workers' compensation, never to restore their lives and their livelihoods again, will depend on the severity of the injury.

Most of us, myself included, thought workers' compensation was there for us if we needed it. How wrong we were! Year after year, with the benign neglect or active collusion of state insurance departments, workers' compensation allows the continuation of a system that kills and maims with impunity.

The title of this book refers to a workers' compensation system in which worker deaths often cost an employer far less than correcting a health/safety hazard in the workplace; and a system in which the employer is virtually immune from prosecution for the depraved indifference that results in severe injury, toxic exposure and death. New York State Penal Law 125.25 (2) defines murder in the second degree, an A-l Felony, "… Under circumstances evincing a depraved indifference to human life, he recklessly engages in conduct which creates a grave risk of death to another person, and thereby causes the death of another person;.…"[3]

The stories in this book are the stories of ordinary people, except that it happened to them. The cushion of workers' compensation they thought they had evaporated, leaving them to their own resources. Nay, worse! Not only was workers' compensation not there to help, except to stuff the insurers' pockets, but it added to the burden to be faced. It forced a battle for medical treatment and for wage compensation, for the very survival of the worker or the bereaved family of the worker who died, at the same time that it accused the injured worker of faking it, the deceased worker of causing it.

With great hoopla, Congress passed legislation to get people on welfare back to work. At the same time, we push hundreds of thousands of injured and ill workers each year from work to welfare and poverty, to poor health with no health care, to broken lives, many never to be able to work or support their families again.

Our nation has turned its back on workers who believed in the system, believed the American dream—right up until the day it all turned upside down.

What has happened to our nation that has allowed corporations and insurers to throw away the lives, the health of so many without caring, without fear of exposure, without being held accountable? It's called Workers' Compensation: the system we thought was there to protect us if we were injured on the job.

Historically, the workers' compensation system began in Germany in the 1880s. Workers' compensation was an opportunity for Bismarck to quell the growing anger over worker injuries and deaths, and at the same time ensure Bismarck's base of power after he had outlawed the trade unions and ordained other repressive "reforms." The concept emigrated to Great Britain and was passed there in 1897. Great Britain, however, began the reform of its workers' compensation system over 50 years ago, recognizing that workers' compensation protected the employer from liability, and thus there was no incentive for employers to make the workplace safe.[4]

In 2002, Great Britain, twenty percent of the US workforce in size, had only four percent of the worker deaths recorded by US BLS in the same year (225 vs. 5,524).[5]

In the United States, the no-fault principle has been enshrined. And that is the core of the problem.

The no-fault principle disenfranchises workers from their Seventh Amendment rights to bring suit at common law for redress of grievances.[6] Under workers' compensation, the worker cannot sue his/her employer for injury or death. The quid pro quo is supposed to be swift and certain compensation and treatment in exchange for having given up the right to sue. What has happened over the years is that the freedom from lawsuit prevails, while worker medical treatment and wage compensation have been delayed and denied away. Thus there is no monetary disincentive for employers and insurers to not pay for the disease that cripples or the corporate negligence that injures or kills. It's in the corporation's best monetary interest not to pay. The longer the money stays in the insurer's pocket and investment portfolio, the more desperate the worker's situation becomes. Thus this system, ostensibly to protect workers, has continued to erode away workers' rights and benefits, without anyone noticing; except employers and the workers' compensation insurance industry that have brought about these changes, and who profit from them mightily. "Starving them out" is the phrase used in the workers' compensation insurance industry, according to a former insurance claims adjuster.[7]

Having lost the right to sue, workers have no leverage in the workers' compensation system; and thus there is no justice for injured, killed, and toxic chemical-exposed workers.

Workers' compensation insurers and their powerful lobbyists, supported by employers, regularly lobby state legislators to shrink even the meager legislated health and wage compensation and to limit worker attorney fees, claiming its too costly for employers. At the same time, average executive compensation has reached a ratio of 431 to one compared to line workers' pay.[8] The injured and the families of the deceased have weak voices, if they are heard at all, in state legislatures. And legislators, who are generally ignorant of the workers' compensation system, and concerned for the pockets of business, acquiesce to the demands of the insurers.

Increasingly, corporations are shifting their responsibilities for worker pensions, for medical care and, for a long time now, for work injuries, occupational illnesses, and deaths onto taxpayers.

One hears a lot about worker fraud, especially at times when the workers' compensation insurance lobby is trying to pass legislation to further reduce the benefits to injured workers, or to restrict access, or compensation to attorneys representing the worker. In this system, often the only thing that stands between the worker and starvation is the worker's attorney. Insurance industry attorneys, by contrast, have no legislative limits imposed on what they are paid.

But who is really committing the fraud? Studies consistently show claimant (worker) fraud at 1% or less. In State of California, state investigators found claimant fraud averaged less than half of one percent over a five-year period.[9] In 1996, State of Texas Research and Oversight Council on Workers' Compensation reportedly found a total of $134,000 in claimant fraud, but $1.2 million in provider fraud.[10] State of Wisconsin found claimant fraud to be less than six tenths of one percent.[11]

What we don't hear about is the insurer fraud, premium fraud (more about the latter in Chapter 6), provider fraud.[12]

The California state auditors are said to have found that half the claims it audited over seven years consistently showed insurers' violation of workers' rights by failure to pay compensation due to workers, as well as late benefit payments, and "materially inaccurate" notices sent to workers. The average fine in 1996 was $126—not much of a deterrent to a large insurance company.[13]

Leigh, et al. (2000) found that, on average, workers' compensation pays only 27% of the $156 billion annual cost of worker injuries and illnesses. Fully 18% of the costs are being paid by taxpayers; through Social Security, Disability, Medicare, Medicaid, public assistance; representing $28.5 billion in 1992 dollars. Almost half the cost, 44%, falls on the injured, ill or deceased workers and their families.[14]

The corporations that cause this grief and pain continue to profit while the injuries, toxic exposures and deaths continue.

The promised wage compensation under workers' compensation, intended to sustain the injured worker and his/her family while unable to work, continues to be shrunk by state legislatures year by year until today the range is from 70% of the poverty level for a family of three (Mississippi) to a mere 170% of poverty (Washington, D.C.) with most states at about 120% of poverty.[15]

All the while, workers' compensation insurers beef up their administrative budgets, raise premiums and take their investment profits to the bank. In State of Florida, over the period 1989–1998, while settlement awards declined, after tax investment profits reached almost $4 billion, in addition to the over 47% of earned premium that insurers paid to themselves.[16] Nationwide, workers' compensation carriers reaped $342 billion in profits by 1998.[17]

Increasingly as a nation, we appear to be all too willing to push corporate costs onto workers and taxpayers; all too willing to cut protections for workers, communities, and the environment.

Safety should be a no-brainer. It saves lives and costs. But if workers' compensation insurers and employers are not paying the preponderance of the costs of worker injuries and occupational illnesses, where is the incentive to improve workplace health and safety?[18]

The federal Occupational Safety and Health Administration (OSHA) has the responsibility for setting safety and health standards, monitoring and ensuring the safety and health of the workplaces. OSHA fines for citations (violations) have been so watered down that it's cheaper for the corporation to pay the fines than it is to provide a safe workplace. In its over thirty years of existence, federal

OSHA reportedly has referred only 151 cases to federal Department of Justice (DOJ) for criminal prosecution, only one third of which were pursued by DOJ, and only eight in which prison time was meted out.[19] Even in those few cases when OSHA deems a citation "willful" and allows it to go to DOJ for prosecution, the crime is considered a mere misdemeanor with a maximum penalty of "a fine of not more than $10,000 or by imprisonment for not more than six months, or by both."[20] Compare that to the criminal fraud and conspiracy charges facing Kenneth Lay, founder and former chair and CEO of Enron that could send him to prison for life.[21] The criminal penalties for the depraved indifference to injuries, illnesses, and toxic exposures that maim and kill workers are miniscule compared to those for cheating investors.

Justice demands that we right this system that now looks the other way when corporations ignore the safety, the health, the very life of workers—all in service to profit. Human life has become expendable in America's workplaces.

Getting an accurate count of workers and former workers who die from occupational diseases in the United States is difficult and complex.

Occupational diseases generally are not clinically different from other diseases from non-occupational causes. The combined effect of occupational and non-occupational effects impacts causation as well.[22] In addition, the long lead time between the workplace exposure and the manifestation of the illness often masks the occupational etiology of the disease occurring years later.[23]

Only a minute number of the chemicals used for commercial purposes have ever been fully evaluated for their potential toxicity.[24] Of the 650,000 hazardous chemicals in use in the workplaces of America, OSHA has only established permissible exposure limits (PELS) for about 300 of them, and most of those PELS are out of date based on more recent available information.[25]

Markowitz, in Leigh, et al.(2000) computed a conservative estimate of the number of illnesses and deaths attributable to occupational causes. This careful and conservative analysis of specific diseases used an attributable proportion risk model to estimate the percentage of disease mortality, within certain age ranges, attributable to occupational disease. An attributable proportion refers to the "proportion of (a) disease that would not have occurred in the absence of that exposure."[26] Over 20 primary and 300 secondary sources were researched in this comprehensive study on the numbers and costs of occupational injuries, illnesses, and deaths.[27] Estimates in the final calculation ranged from one percent to ten percent of specific selected disease deaths, plus 100% of pneumoconiosis deaths. For 1992, the total worker death estimate from occupational diseases in the United States was 60,293 persons.[28]

Comparing the estimated occupational disease deaths 1992 to recorded US deaths from all causes in 2002, and conservatively assuming no increase over the years, occupational disease deaths would still be more than breast cancer deaths; over twice the number of prostate cancer deaths; more than deaths from colon and rectal cancer, Alzheimer's disease, heart failure; about three times the number of deaths from leukemia; four times the deaths from HIV. Yet, all of these diseases have received far more attention than occupational diseases.[29]

For the reasons just described, worker deaths that do not occur in the same year as the injury or toxic exposure that caused them, would not likely be included in the US Department of Labor, Bureau of Labor Statistics (BLS) data. BLS is, however, the most comprehensive source on nonfatal work injuries, although some categories of workers are excluded altogether, e.g., government workers, postal workers, farms with less than eleven employees, private household employees; and some businesses don't report or underreport, even though it's illegal. As already noted, occupational toxic exposure illnesses are not likely to be counted in the BLS figures.

Glazner, et al. (1998) cited in Leigh, et al., identified significant underreporting of worker injuries; 20% for disabling injuries, and 35% for non-disabling ones. There will be more discussion on underreporting in Chapter 8, Medical Treatment under Workers' Compensation.[30] Leigh, et al.(2004) found even higher undercounts of 33% to 69%, when those worker categories excluded from the BLS survey were factored in.[31]

Everyone whose story is in this book had a life, family, people whom they loved, and who loved them. Those who lived found their lives changed forever—financially, physically, emotionally. Those who died left behind families, memories, and the pain of how they died.

These are the stories of the lives of just a few of the millions of workers whose lives and health are sacrificed on the altar of corporate greed every year.

For those who have been occupationally toxic chemical-exposed, the worker is often the "canary in the coal mine:" The family, the children, the aged, the community all suffer from the toxins that industry is releasing into the air, water, and soil.

When workers' compensation cases are closed, my own included, the case is generally settled with the proviso that neither the employer nor the settlement information is to be disclosed. Thus, the documentation in this book varies from chapter to chapter. Cases in which death has occurred have state or federal Occupational Safety and Health Administration (OSHA) investigations and other documentation, as a rule. In Dale Goldstein's case, his family shared with me the journal that he had kept while he was dying.

In those cases in which either the case was sealed or I was unable to locate family in order to get permission, I have used a fictional name and identified it as such in the chapter endnote. In such cases, any similarity between the fictitious names and the names of any persons living or dead is purely coincidental.

Human destruction on such a monumental national scale deserves national attention. For too long, the scales of justice have been skewed to the insurer/ employer side, while whittling away at the meager protections for injured and toxic-exposed workers, and the families of those who died. Justice delayed is justice denied. Workers have been terribly wronged and left without the right to adequate redress that the constitution guaranteed to all. Right now, an employer would have to put being a decent human being above profit in order to fix his/her workplace. It doesn't have to be like that.

The last chapter explores where we can go from here to create a system of fairness and justice, a system that works, that puts the responsibility and accountability on the employer to make the workplace safe and healthy for workers.

References:

1. United States Department of Labor, Bureau of Labor Statistics, www.bls.gov.

2. Leigh, J. Paul; Markowitz, Steven; Fahs, Marianne; Landrigan, Philip. *Costs of Occupational Injuries and Illnesses.* University of Michigan Press, 2000.

3. New York State Penal Law, 125.25 (2) Murder in the second degree (A-1 Felony).

4. White, Lawrence. *Human Debris: The Injured Worker in America.* Seaview/ Putnam, New York. 1983.

5. Thou Shalt Not Corporately Kill. The Economist, May 24, 2003, p. 59 and BLS data, cited in A National and State-By-State Profile of Worker Safety and Health in the United States, *Death on the Job: The Toll of Neglect,* 13th Edition, AFL-CIO, Washington, D.C.. April 2004.

6. Bill of Rights, The first ten amendments to the Constitution of the United States, ratified December 15, 1791. Amendment VII, which reads: "In suits at common law, where the value in controversy shall exceed twenty dollars,

the right of trial by jury shall be preserved, and no fact tried by a jury shall be otherwise re-examined in any Court of the United States, than according to the rules of common law.

7. Workers comp: Falling down on the job. Consumer Reports, February, 2000, p.29.

8. Study by Institute for Policy Studies, cited in "Too Many Turkeys." The Economist, November 26, 2005, p.75.

9. Fricker, Mary. Widespread fraud: Bogus claim. Santa Rosa Press Democrat, at pressdemo.com/workerscomp/dayl/fraud2.html, 1997.

10. Law Makers to discuss Workers' Compensation Issues, January 23, 2002 in Tallahassee, Florida. State Compensation Insurance Fund, p.2. Chronic Pain Foundation at www.chronicpainfoundation.org/WorkersComp.html.

11. State of Wisconsin, Department of Workforce Development, "Workers' Compensation Fraud is Low," Press Release, November 26, 1997, cited in Cullen, Lisa. The Myth of Workers' Compensation Fraud, Frontline, 2003, at www.pbs.org/wgbh/pages/frontline/shows/workplace/etc/fraud.html.

12. O'Malley, Todd J. Fighting Insurance Fraud For Your Client's Sake. Reprinted from Trial Magazine, December, 1998. Association of Trial Lawyers of America,www.omalleylangan.com/pub2fighting.html.

13. Fricker, Mary. State audits find insurers routinely unfair to workers. Santa Rosa Press Democrat, 1997 at www.pressdemo.com/workerscomp/dayl/audit/html.

14. Op. cit. Leigh, et al.,2000, pp. 1, 3, 11, 175.

15. Hunt, A., et al. Adequacy of Earnings Replacement in Workers' Compensation Programs, unpublished study of National Academy of Social Insurance, Washington, D.C. 2004, cited in op. cit. AFL-CIO, *Death on the Job: the Toll of Neglect.*

16. Jordan, A.D. II. A Study of the Profitability of Workers' Compensation Carriers in the State of Florida: By Analysis of Carrier Annual Statements Filed With The Florida Department Of Insurance, 2001, p. 4.

17. Slip Sliding Ahead. A M Best Review, January, 1998, cited in Workers' Compensation Reform: Why Is It Needed? Boxer and Gerson, 2000 @ boxerlaw.com/bg04024b.htm.

18. Cullen, Lisa. Safety Pays—Or Does It? The Synergist, April, 2003.

19. Criminal Prosecutions of Workplace Fatalities. Frontline, WPBT, 2003 @ pbs.org/wgbh/pages/frontline/shows/workplace/osha/refe rrals.html.

20. OSHA Field Inspection Reference Manual CPL 2.103, Section 4, Chapter III, C.2.e.

21. Mcclam, Erin. Lay: Legacy a source of pain. The Associated Press, in South Florida Sun-Sentinel, p. 1A, 4/25/06.

22. Op. cit. Leigh, et al., 2000, p.56.

23. ibid, p.57.

24. ibid, p.57.

25. Cullen, Lisa. *A Job To Die For.* Common Courage Press, Monroe, ME, 2002, p.75.

26. Markowitz, in Leigh, et al.(2000), Chapter 3, pp. 55–89.

27. Op. cit. Leigh, et al.(2000), p.3.

28. ibid, Markowitz, Chapter 3, Number of Illnesses, Table 3.12, Estimated Occupational Disease Mortality Due to Selected Causes, United States, 1992, p.87.

29. The data source is the US Department of Health and Human Services, National Center for Health Statistics, Centers for Disease Control and Prevention, National Vital Statistics System, National Vital Statistics Reports, Vol 53, Number 5. Deaths: Final Data for 2002, Table 10 and Worktable I, pp. 1585, 1634, 1662, 1703, 2220–2224, at cdc.gov/hchs/data/dvs/mortf inal2002_worklpt2.pdf.

30. Glazner, J.F.; Borgerding, J.A.; Lowery, J.T.; Bondy, J.; Mueller, K.L.; and Kreiss K. 1998. Construction injury rate may exceed national estimates: Evi-

dence from the construction of Denver International Airport. American Journal of Industrial Medicine 34:105–112, cited in Leigh, et al. (2000) op cit.

31. Leigh, J. Paul; Marcin, James P.; Miller, Ted R. An Estimate of the U.S. Government's Undercount of Nonfatal Occupational Injuries. Journal of Occupational and Environmental Medicine, 46 (l):10–18, January 2004.

CHAPTER 1

▼

DALE ALAN GOLDSTEIN

At 47 years of age, Dale Alan Goldstein was a 6' tall, physically fit, 180-pound man with blue-gray eyes and a light brown receding hairline that complimented his Tom Selleck-like moustache. A divorced father of four, with two grown children and two still at home, he was a loving father to his children and a homebody who loved spending time and doing things with his kids. His youngest son, Brandon, idolized his father.

For Dale's memorial service following the cremation of his remains, his son Brandon, aged 15 years, wrote the eulogy entitled "My Lost Friend." The Eulogy follows:

"I have lost something precious today. Not something made of gems or material objects, but rather a gift only from God. This gift was my father. But he wasn't just a great dad, he led me to the lord, he encouraged me when I was down. He did this not for his own gratification, but because he wanted to see me happy and wanted me to succeed. My dad was someone who wanted to spend time with me, and that made me his friend. Often today parents and children grow apart because of their hectic lives. My friend would always have time for me. We would make spectacular dinners and we took trips around the state. I remember he stayed up all night with me just to catch a couple of freight trains go by. And when I sensed that he was tired we would always stay a little longer to see the next train. He truly is my friend. He did not treat me as a boy, but as a man and an equal. This helped us grow closer together and we were able to talk about anything on our minds. I encourage the parents and children to spend time together because we often take each other for granted. I was at least fortunate to spend time with my dad in his last few years. I thank you, Dad, for your love.

"My dad has a free spirit. I remember one night he came over and asked me if I wanted to take a road trip to Georgia. I was caught by the sudden surprise, but was excited to go on the trip. So, shortly after that we were heading for Georgia enjoying the cool night air. On the way there he said, "Life is just more exciting not having a schedule to follow. You never know what to expect." I agree with what he said and I think the element of surprise is exciting. We had an adventurous life and we enjoyed every minute we spent together. It didn't matter if we were cooking, camping, train spotting or just sitting around a campfire, we were happy to be together.

"My dad has love that will never die. He will always look down on us from heaven. He never hesitated to say that he loved me. I had recently looked at a picture of my dad and myself in it. It wasn't the first time I had seen the picture, but I noticed something different about it this time. It was my dad's eyes. They were not focused on the camera but rather affixed to the smile on my face. His face was gleaming with expressions of love and joy and was glad to spend his precious time with me. Whenever I look at that picture, it is reassurance that he has loved me since I was in my mother's womb and will love me through eternity. He would bring me comfort with his radiant glow as if he were an angel sent from heaven.

"But now he is with the Lord and perhaps he is the angel that I have always envisioned him to be. Not one of us ever dies. It is our flesh that dies but our spirits live on forever. If we have our spirits then we have our souls and ourselves. But we miss the flesh because that is what we touch, see and feel. Be comforted to know that God has now touched his spirit and the spirit is much greater than the flesh. He is at peace now, and his sickness healed by the Almighty one. So never forget that my dad is now with our Father and he lives on in heaven.

"My dad is a creation of God that we will surely miss and he has moved on to a place that is apart from sin and suffering. It's kind of ironic that we leave this earth as in which we came: from dust to dust once again. God brought life with dust and we leave as dust. Perhaps the major difference is we leave with a heart and a spiritual body. And I thank God for our lives. I am relieved that it is not the end of time, but nearly the beginning. I write this in remembrance of my dad and to comfort my family and friends and to help me through my grief. And remember one day we will see our Lord and we will see my dad in joyful occasion."

*   *   *   *

From age eighteen, Dale Goldstein had been an executive in his father's chain of men's clothing stores. He had also studied for the ministry. Over the years, he and his wife had established their own home-based, rather successful business. With the divorce, they decided it was better if she continued with the business and he worked elsewhere.

As he soon discovered, despite his broad experience in business and his time in theological seminary, his lack of a college degree was a limiting factor in his job search.

Dale began working through a staffing company. Some positions were exhilarating, e.g. he worked on a zeppelin marketing tour for a major candy manufacturer. In his letters home during this time, Dale describes traveling from city to city and what the work entails. The ship was on a steel tower mast about 20 feet high, anchored with steel cables. Hourly readings of pressure, air temperature, weather changes, weight adjustments and other calculations were necessary when on the mast. He describes the challenge of the weather and describes a day when they were "in full throttle and going backwards!" Dale explains the zeppelin could go 25–30 miles per hour, thus a 30-mile per hour headwind would propel it backwards. "I enjoy the people who visit the ship. We answer all their questions, put them on board and give them sweets, compliments of the company."

His next position through the staffing company was for a maintenance contracting company, for which Dale was to do air conditioning preventative maintenance in several states. He was to perform contract work, remaining in a particular state for 28 days, using company-provided pickup truck, equipment and supplies. He started this work in October 1998.

The company Director of Safety provided a few hours of initial training for new service personnel, or as Dale describes it, "an industry video on the principles of refrigeration." There was no information provided on the chemical, sodium hydroxide, that he would be using. So Dale queried the Safety Director after the training session:

Dale: "What if I get some on me?"
Safety Director: "You just rinse it off", said with a light-hearted smile and hand gesture.
Dale: "Does it burn you? Is it acid?"

Safety Director: "Oh, no. It's not an acid. It's a base. You might feel a little tingle." He smiled.

Dale adds, "I was comforted. His parting joke altogether allayed any concerns that remained in me. 'Just don't drink it,' he said. I had nothing to worry about."
1

As it would turn out, what Dale didn't know would kill him. Dale Alan Goldstein died of Systemic Lupus Erythematosis (SLE). Two weeks of exposure by inhalation, skin and eye contact to high concentrations of sodium hydroxide, eight hours daily, without proper protective equipment, triggered the body systems breakdown that took place over the ensuing 11 months, ending in the final collapse of all body systems—death.

His death was slow and laborious. "Systemic" is such a sanitized word: It doesn't begin to capture the full essence of what it means when every major body system breaks down, painfully, horribly. In Dale's case, deteriorating so completely that within 11 months from his first exposure, this healthy 47-year-old man was dead.

How did this happen? Why did it happen? Dale became a chronicler of his own demise, writing his story as it was unfolding.

"In Kentucky, I was instructed by a field trainer who introduced me to the equipment I would be using and demonstrated the procedures. I was issued a safety harness, goggles, disposable paper masks for my mouth and nose, and the sodium hydroxide (NaOH).

"I was shown how to mix the NaOH in a garden type insecticide sprayer, mixing it with water in a 2:1 ratio of water to sodium hydroxide. The sprayer wand was about 24" long with a spring handle, which opened the valve to release the chemical, and the sprayer nozzle where the chemical exited had a 2 or 3 piece finger-knurl apparatus, which needed continual adjustment to optimize application of the chemical. When you were spraying down the coils within the rooftop AC units, which were at the farthest reach from your open access panel, you would hope your arm plus the sprayer arm … would reach the coils … six feet away. This would necessitate finger contact with the wet nozzle dozens of times each day. The only gloves issued were super heavy impervious type which were only used for the mixing…. and of no use for nozzle adjustments, as they prohibited any fine motor dexterity due to thickness.

"My head was a constant target of the vapor/spray/mist. My face and hair were being continually sprayed as the convection, arising from the store's interior ceiling would naturally exit upward into the rooftop return-air duct above which

I was working. I was invariably spraying into the wind as the rising/exiting store air carried the vapors toward me as they would escape via the open panel. My goggles were in continual need of a rinse due to the mist.

"Within a few days I came to realize that this mist was coming through the gaps around my goggles and the paper mask which I was wearing over my mouth and nose. I found that when I put the mask over my nose and mouth just right, the top of it interfered with the goggle seal at the bridge of my nose. Conversely, if I gave preference to the goggles, the paper mask would have a considerable gap. This brought my chemically-wet hands to my face for adjustments no less than a couple of times per hour.... I was not unaware of the continual mist that I found myself ingesting by mouth. Upon the first warm weather day, I found my arms continually burning and soon realized it was the first time I had no sleeves on while I worked.... The more careful I became and began to rinse my face more often."

At the end of the first week, after about 60 hours of contact, Dale writes, "I found my throat and lungs parched. When I laid down to sleep at night, ... woke continuously to quench this strange dryness, drinking 16–20 ounces of water a night. Within a few days, ... by end of the first seven days spraying, I requested a full-face mask from the company president. I was told ... not to buy a new mask ..., to await the head of field training, who would be sure I got the right one...."

"Each day after work, I drove to the next town so that I would be ready for the next day's work. To meet the head of field training in Indianapolis, my trip after work that day was several hours, longer than usual. That night, I was so tired, I just went to bed without a shower. The next morning after my shower, I was shocked to find a clump of hair at the drain!"

When Dale met with the head of field training, he reported his concerns and requested a full-face mask for his use. The head of field training told Dale the company wanted him to see a specialist, but none could be scheduled for three weeks.

Dale then called the manufacturer of the mask and asked whether that was the correct one for sodium hydroxide. "The answer was immediate, 'Not for sodium hydroxide. You need a full face respirator, one with a 100 rating ... or a powered air mask,'" he was told.

National Institute of Occupational Safety and Health, (NIOSH), Dale learned, "has a standard which cannot be met on rooftops with open winds and uncontrolled store convection misting your face, short of the full face/powered air respirator that ... recommended". In fact, the manufacturer's standard had come from NIOSH.

Dale then went back to the head of field training, who "did not have knowledge of the vapor levels in which the work was done."

In November, 1998, the first of the trips to clinics, and hospitals began. A walk-in clinic in Indiana gave him something that would moisturize his lungs and throat. The company told Dale to return to Orlando, Florida to see a specialist. "The specialist ... was also found to be weeks away, so I visited another walk-in clinic, was given a bronchitis type prescription and told to come back 5 days later, November 5, 1998. I was then classified as 'light duty,' for which there was no work available. I was by now clear of symptoms, having been away from work for 7 days. I repeated my request for the proper mask and asked to be released back to work. I was checked off as 'resolved.'"

It was during this time that Dale ordered the NIOSH manual on sodium hydroxide, 95 pages, according to Dale. By the time the manual arrived, shortly before Thanksgiving, Dale was no longer employed by the company. His primary symptoms having waned, Dale "spent only an hour reading of sodium hydroxide's power, glad only that that nightmare was over—all the while it hadn't even begun."

Dale's mother, who saw him that Thanksgiving, described him as looking weak and exhausted. One month later, by Christmas, she says Dale was "all bloated, and one eye was practically closed."

What is sodium hydroxide (lye) and what does it do the body? Dale does his own research and finds:

- Sodium hydroxide is "a corrosive irritant to skin, eyes and mucous membranes.... Inhalation of the dust or concentrated mist can cause damage to the upper respiratory tract and to lung tissue, ... has a markedly corrosive action upon all body tissue (,) ... causing burns and frequently deep maceration...." It is generally thought to be more damaging than strong acid.[2]

- The lethal oral dose of caustic alkalis in humans is less than 10 grams.[3]

- Exposure routes include "inhalation, ingestion, skin and/or eye contact."[4]

- "The severity of injury depends on the concentration of the solution and the duration of exposure."[5]

- "Severe burns with deep ulcerations will occur" if the tissue is not immediately decontaminated.[6]

- "Absence of distress marks latent period and that esophageal strictures will develop within weeks or months unless effective treatment is instituted."[7]

- Ingestion of sodium hydroxide causes "corrosive injury to the mouth, the throat, esophagus, and stomach … and may result in perforation, hemorrhage, and narrowing of the gastrointestinal tract."[8]

- "Alkali burns … can be underestimated as to their severity, based on initial observations. Sometimes they are latent in their onset, due to formation of alkaline albuminates and fatty soaps in the tissue."[9]

- "Contact with the eyes causes disintegration and sloughing of conjunctival and corneal epithelium, corneal opacification, marked edema, and ulceration. After 7 to 13 days either gradual recovery begins, or there is progression of ulceration and corneal opacification. Complications of severe eye burns are symblepharon(adhesion of the lid to the eyeball) with overgrowth of the cornea by a vascularized membrane, progressive or recurrent corneal ulceration, and permanent corneal opacification."[10]

NIOSH recommends

1. Comprehensive pre-placement medical examinations for any workers subject to sodium hydroxide exposure.

2. "Medical examinations shall be made available promptly to all workers with signs or symptoms of skin, eye or upper respiratory tract irritation resulting from exposure to sodium hydroxide."[11]
   Neither of these was done in Dale's case.

The degree of exposure, the concentration of the solution, and the susceptibility to the effects for each individual seem to determine the severity of the disease. Dale states, "there is little known data for organ damage at high exposure levels such as I incurred. Nonetheless, the Poison Control Center of Florida recognizes the danger posed to the kidneys by (sodium hydroxide) and recommends renal testing upon ingestion."

As Dale points out, most sodium hydroxide exposure is occurring in industrial settings in which exposure levels can be easily monitored. NIOSH recommendations include fresh air ventilation and exhaust systems. Dale writes, "This controlled environment being the case with most use, again would indicate very little data available, very few high exposure injury records with which to establish typical kidney responses."

The exposure which Dale described in his chronicle thus far, was not, however, the whole picture. As he began his reading on sodium hydroxide, he realized that even "the bed liner of the truck was coated with a white powder, the metal ladder as well, and many other surfaces with which I was in constant contact. I had believed the powder was the effect of the chemical; not what I now know to be the chemical itself, upon evaporation. The powder form is considered equally as dangerous; ... so when I climbed into the rear of the covered pick-up for my laundry detergent, replacement fan belts, or my cooking grill, I was often in contact with the chemical itself, and in ignorance, did not rinse it off."

Material Safety Data Sheets (MSDS) have been made available to workers for many years. The aim is to provide a safer working environment by providing the information a worker needs to know about a particular substance.

The Material Safety Data Sheet (MSDS) issued by the manufacturer describes the product Dale used as "Corrosive to all body tissues."[12] The MSDS also identifies potential health effects: Inhalation of vapor or mist is corrosive to respiratory tract, liquid or vapors are corrosive to eyes, and skin, and it lists a chronic effect, chemical pneumonitis.[13] With regard to handling, it warns, "Avoid contact with skin, eyes and clothing. After handling this product, wash hands before eating, drinking or smoking. If contact occurs, remove contaminated clothing. If needed, take first aid action.... Launder contaminated clothing before reuse."[14]

MSDS Section VIII Exposure Controls/Personal Protection states "Local exhaust recommended." There was none where Dale was working; in fact, the wind and the AC system were blowing the concentrate back into his face and skin. Section VIII also states, "Wear chemical impervious gloves and full face shield. Use boots, aprons, drench showers, eye wash as needed for protection against spills and/or splashes."[15]

None of this, not full-face shield, nor boots, nor aprons, nor eye-wash was provided to Dale. Nothing, except possibly gloves, met the standard and was made available to Dale. And the employer never gave Dale the MSDS. Dale called the manufacturer and got it from them.

"When caustic soda" (sodium hydroxide)"comes into contact with the skin, it does not usually cause immediate pain, but it does start to cause immediate damage. It fails to coagulate protein, which would serve to prevent further penetration.... Following contact with skin, washing with water must be started immediately to prevent corrosive chemical burns."[16]

Note that the MSDS requires "full face shield" and that what the company issued was a disposable paper mouth and nose mask and separate eye goggles. Just a few months before Dale was hired, the company issued an internal memo to All Service Personnel from the Safety Officer, dated July 17, 1998 that required the use of a disposable mask and goggles as mandatory equipment. But it was the wrong safety equipment, insufficient and inadequate.

OSHA's permissible exposure limit (PEL) for sodium hydroxide is 2mg/m3 per 8 hours.[17] OSHA itself concludes that this is too high: that a ceiling of 2mg/m3 is "necessary to reduce the significant risks of eye and skin burns and respiratory irritation that occur as a result of very brief exposures to the higher levels of sodium hydroxide that would be permitted with an 8 hour time—weighted average (TWA) PEL alone. OSHA considers the irritant effects ... material impairments of health."[18]

The National Institute of Occupational Safety and Health (NIOSH) recommends a 15-minute exposure limit at the ceiling value of 2 mg/cu m.[19]

OSHA has been under-funded and embattled by industries, which see profits as more important than human life and fight every attempt at enhanced PELS to the death. And the battle has succeeded. As noted earlier, of the 650,000 chemicals used in industry, only a miniscule number have had PELs established in the almost 30 years of OSHA's existence, and even those have not been upgraded with advanced information. In other words, even in the small number of cases where OSHA has established PELs, these too are inadequate and ineffective to protect workers from disease and death.[20]

Dale Goldstein's exposure to NaOH was approximately two weeks during October 1998. By November, 1998, Dale describes "congestion, coughing, wheezing when reclined, low grade fever ... ever present daily, diarrhea, abdominal pain, eyes discharge and tearing." He assumed he was fighting the flu.

By December, all these symptoms continued. Dale's journal reads, "eyes and fever worsened, legs began to swell up and strength declined sharply. Sleep increased from the normal 7–8 hours per night to 9–10 hours minimum, with 5–7 hours additional needed in naps."

Dale says, "I had only one doctor visit the entire decade aside from hernia repair, check-ups," referring to his pre-exposure health status.

On January 5, 1999 he was "diagnosed with pneumonia, conjunctivitis, arthritis." Dale was "instructed to check into a hospital, but I didn't due to lack of insurance." But he did insist that some tests be run, which was done that weekend. On January 12, 1999 he checked back with the physician for the test results, and was told to return and "get medical care at once" or check into the hospital in the town where he was working. At this time Dale was working for another company. Dale writes, "urine and blood results showed serious aberrations from norm. Also, my visit on January 5th showed high blood pressure: I'd always been in perfect if not low range. I tried to resume work, but chest pain prompted me to call ambulance."

On January 18, 1999, Dale consulted with a nephrologist and was transferred to a regional medical center for kidney biopsy. "Preliminary diagnosis was nephritic syndrome, biopsy further indicated SLE (systemic lupus erythematosis), lupus as highly likely…. Proteinuria severe, anemia, edema, pneumonia…. Cardiologist found fluid sac on heart, thickened wall…. Weight gain of 50 pounds in a couple of weeks (175 to 225 lbs). No strength." Dale was released from the hospital January 23, 1999.

Dale returned to the hospital again on March 10, 1999 for 2 units of blood. In a "five day stay, shed 25 pounds on IV diuretics," chronicles Dale. He writes, "January and February were consumed by the sickness itself and dealing with the reality of an incurable disease, which is now assailing its fifth organ; along with a host of smaller problems, like legs that don't want to climb stairs, hypersalivation, thrush, insomnia, sub-retinal edema.

"In my initial studies of lupus," Dale writes, "I found that etiology was related perhaps as much to environment as it was to genetic predisposition. In my inquiries to the … Poison Control Center, (FL), the Director of Medical Research of the National Kidney Foundation, a Medical Director from the New York University School of Medicine and the plethora of research forwarded me by the Division of Toxicology of National Institute of Health (NIH), Centers for Disease Control (CDC), NIOSH; I have found a solid foundation upon which rational medical opinion is based for a causal relationship between the toxic exposure herein described with the concomitant symptoms experienced…. At issue here is not whether or not a nephrotoxin can cause lupus. Rather, did it accelerate its dormancy to the extent to bring a 47 year old man in the physical condition of top 10% of men his age into a totally disabled state in a matter of weeks."

The National Institute of Environmental Health Sciences September 1998 workshop posed the question of "whether environmental agents affect the development or progression of autoimmune disease." It concluded that there were no well-validated methods for assessing autoimmunity at that time, although there is "evidence for the role of environmental agents in the initiation or progression of autoimmune conditions."[21]

A hospital in Denmark describes irreversible obstructive lung disease after a healthy 25-year-old man has worked with sodium hydroxide for one day in a poorly ventilated room.[22]

In a study of a small community exposed to a number of industrial toxins, an extraordinarily high incidence and prevalence of systemic lupus erythematosus is identified. The authors find that "the hypothesis that environmental toxins may induce lupus is consistent with the known ability of certain medications to do the same.... Long-standing exposure to industrial emissions may be associated with an increased risk of lupus."[23]

"The research," Dale continues, provides the "needed framework to establish sodium hydroxide's role in escalating an innocuous condition into one which is life threatening.... (,) to have caused most every injury directly and without a relationship with a preexisting factor."

Dale poses the paramount question: "what are the extraordinary odds that all of these organs were assaulted coincidentally, in perfect timing with a gross exposure to a most deadly toxin?"

What an extraordinary man was Dale Goldstein, to do the research so analytically on his own disease. At the same time, he has to do battle to justify that his condition was caused by the chemical to which he was exposed by his employer. Even if this had been an aggravation of a preexisting condition, workers' compensation in State of Florida,(the state of the employer) allows for proportionate compensation for a work-related aggravation of a preexisting condition, causing death or a compensable permanent impairment.[24]

OSHA, too, recognizes this in its Standards on Determination of Work-relatedness: "you must consider an injury or illness to be work-related if an event or exposure in the work environment either caused or contributed to the resulting condition or significantly aggravated a pre-existing injury or illness."[25]

This author could find no evidence that Dale Goldstein's extreme toxic exposure and subsequent death was ever reported to OSHA, despite being required by law.

In his "Conclusion" which appears to have been written in March 1999, Dale writes, "My knowledge presented here is very much in the order in which I acquired it. I learned about lupus, and it led me to etiology, which led me to environment, which led me back to exposure, which led me to the study of nephrotoxins."

In April 1999, the Florida State Division of Vocational Rehabilitation determines Dale's "disability is to (sic) severe at this time for rehabilitation services to result in an employment outcome."[26]

In June 1999, Social Security Administration declares Dale disabled effective January 1999, and informs him that he will begin receiving monthly checks in August.[27]

In an effort to achieve some semblance of normalcy while he is confined to a nursing home awaiting surgery, he takes a bus to shop so that he can prepare meals on the barbeque grill in the yard at the nursing home: A meal for his children. One of Dale's major enjoyments had been cooking "the spectacular meal," as he called it, for his children. Dale had done all the cooking at home.

By June 1999, Dale is again hospitalized, airlifted in critical condition; so that he could be managed by a pancreatic specialist. He has continuous nephrosis and acute pancreatitis. There is internal bleeding and pseudo-cyst formation. During this time, Dale has exploratory surgery. "They tried to remove his spleen, everything was stuck together and so deteriorated that they couldn't find what they were looking for," tells his mother.

From this point until his death in September 1999, Dale's life is confined to the hospital intensive care unit, on a respirator, on dialysis, on chemotherapy, on a feeding tube, and undergoing exploratory surgery. Dale is unable to speak and can communicate only through gestures and facial expressions.

The hospital is located in the city where his mother lives. Dale was "coming home to die," she says, "and I was the one to see him through his last 90 days."

"A child is not supposed to predecease a parent," Dale's mother says as she tells how she spent the last three months of her son's life. "You watch in disbelief as the life goes out of your child, helpless to make right a terrible wrong.

One day, he looked at me and gently put his hand on my cheek." His mother remembers this very loving moment of tenderness with tears in her eyes.

"My son will never know the joy of holding his beautiful little granddaughter in his arms. As I watched this enchanting baby bouncing up and down in her car seat at the sight of a new storybook of ladybugs and such, my heart filled with longing that I could never share such a moment, ever again, with my dear son," writes his mother.

"Autopsy confirmed the clinical findings. Almost the entire peritoneal cavity was bound down in adhesions and loculated pseudocysts to the level of the pelvis ...

"Additional findings included adult respiratory distress syndrome and focal pneumonia, an abscess of the liver and adrenal cortical medullary hematomas.

"There were mesangial abnormalities of the kidneys, but classical changes of lupus erythematosus, in the kidneys, or elsewhere, were not evident.

"The sequence of events after the NaOH exposure, described by the decedent, in a narrative summary, is compelling, and it seems possible, to me, that this exposure initiated or triggered the subsequent development of his ultimately fatal disease." ... But the Medical Examiner adds, he can't substantiate this from the research.[28]

"The Cause of Death is Sequelae of Systemic Lupus Erythematosus."[29]

To read Dale's chronicle of the series of events tears your heart out; knowing what he does not know, that he will die a horrible and painful death. Having done his job, he was a throw-away person to his employer.

Dale Alan Goldstein died because of toxic inhalation and exposure to sodium hydroxide. The inhalation and exposure to sodium hydroxide occurred because Dale was not given the protective equipment, materials, and information he should have had:

1. He was not told the dangers of the substance with which he would be working, nor did he receive the material safety data sheet (MSDS) on that substance, sodium hydroxide.

2. The protective equipment provided was grossly inadequate. Instead of a carefully fitted full-face respirator with high efficiency particulate filters or a powered air-purifying respirator with dust and mist filters that he should have had, the company provided a disposable paper mask at a much cheaper cost to the company, $17.48 for 20 masks, as noted in an internal company memo. Dale should have been given a complete covering garment with total skin protection. Each time he left a job, the outer protective garment needed to be placed in a sealed container. He should have been issued protective garments in sufficient quantity for his work-days in the field, as well as the sealed container in which to put contaminated garments. Dale received none of this. Nor had he been given eyewash with which to decontaminate his eyes. The truck itself was a further source of contamination. And finally, he wasn't able to decontaminate immediately upon contact.

3. Information on what to do should he inhale or make skin contact with sodium hydroxide should have been provided. In fact, Dale was told that it wasn't harmful at all.

4. Swift, effective treatment should have been made available to him. It was not. When Dale raised concerns, the company scheduled an appointment three weeks away.

Dale was left without medical coverage, in violation of even the meager protections workers have under workers' compensation in State of Florida. The workers' compensation system failed Dale Goldstein completely.

In his "Conclusion," Dale describes how he has been placed in the defensive position of explaining why it took him so long to identify his illness. This man died because of the company's careless and utterly callous disregard for his health and life. The dangers of sodium hydroxide should have been well known to the company and its safety director.

And in the end, the workers' compensation insurer paid the family of Dale Alan Goldstein, father of four, two underage children, the sum total of $5000. As his mother says, "Surely a life is worth more than this paltry sum."

By contrast, the family of an unmarried recent college graduate, who worked at the World Trade Center and died in the September 11th bombing, received over $1 million dollars.[30] This family, of course, should be compensated for the loss of their loved one. But the contrast with what routinely happens to the family of a worker who dies on or from the job is startling.

References

1. Dale Goldstein kept a journal throughout most of his illness, until he was too disabled to do so. The quotes attributed to Dale throughout this chapter are from that journal.

2. Lewis,R.J.: *Sax's Dangerous Properties of Industrial Materials.* 9th ed. Volumes 1–3. New York, NY: Van Nostrand Reinhold, 1996. p.2970, cited in Hazardous Substances Data Bank (HSDB), US National Library of Medicine, and Canadian Centre for Occupational Health and Safety, Issue 99–3, August, 1999.

3. Gosselin, R.E., Smith, R.P., Hodge, H.C.: *Clinical Toxicology of Commercial Products.* 5th ed. Baltimore: Williams and Wilkins, 1984 p. III-246 cited in Hazardous Substances Data Bank, US National Library of Medicine, Bethesda, MD, updated November 15, 2002.

4. NIOSH Pocket Guide to Chemical Hazards, Sodium Hydroxide, CAS 1310–73–2, National Institute for Occupational Safety and Health, Centers for Disease Control, www.cdc.gov/niosh/npg/npgd0565.html.

5. Health Effects of Sodium Hydroxide Solutions. Canadian Centre for Occupational Health and Safety, January 1998 at www.ccohs.com/oshanswers/chemicals/chem_profiles/sodiu m_hydroxide/health_sod.html.

6. Mackison, F.W., Stricoff, R.S., Partridge, L.J., Jr.(eds.)NIOSH/OSHA— Occupational Health Guidelines for Chemical Hazards. DHHS (NIOSH))

Publication no. 81–123 (3 VOL.5) Washington, DC: US Government Printing Office, Jan 1981.2.

7. Gosselin, R.E., Smith, R.P., Hodge, H.C.: *Clinical Toxicology of Commercial Products.* 5th ed. Baltimore: Williams and Wilkins, 1984 p. III-246 cited in Sodium Hydroxide. 5.0 Toxicity/Biomedical effects. Hazardous Substances Data Bank, US National Library of Medicine, Bethesda, MD, www.tox-net.nlm.nih.gov.

8. ToxFAQs for Sodium Hydroxide, CAS 1310–73–2, Agency for Toxic Substances and Disease Registry (ATSDR), www.atsdr.cdc.gov/tfacts178.html.

9. O'Donoghue, J.M., Alghazal, S.K., McCann, J.J.: Caustic soda burns to the extremities—difficulties in management. Br J Clin Pract 1996; 50:108–110 as cited in Sodium Hydroxide (HAZARDTEXT Hazard Management). In: Klasco RK (Ed): TOMES System. Thomson Micromedix, Greenwood Village, Colorado, 2000.

10. Mackison, F.W., Stricoff, R.S., Partridge, L.J., Jr.(eds.)NIOSH/OSHA— Occupational Health Guidelines for Chemical Hazards. DHHS (NIOSH)) Publication no. 81–123 (3 VOL.5) Washington, DC: US Government Printing Office, Jan 1981.2.

11. Sittig, M. *Handbook of Toxic and Hazardous Chemicals,* p.606, 1981, cited on www.toxnet.nlm.nih.gov.

12. Material Safety Data Sheet, Alki-Foam, Virginia KMP Corporation, Dallas, Texas, revised 9/27/96.

13. ibid.

14. ibid.

15. ibid.

16. Kirk-Othmer. *Encyclopedia of Chemical Technology, 3rd ed.,* Vols 1–26. New York, New York. John Wiley & Sons.1978–84, p. 1:861, as cited in Hazardous Substances Data Bank (HSDB), US National Library of Medicine, and Canadian Centre for Occupational Health and Safety, Issue 99–3, August, 1999.

17. 29 CFR 1910.1000

18. OSHA comments from the January 19, 1989 Final Rule on Air Contaminants Project extracted from 54FR2332 et. seq. This rule was remanded by the U.S. Circuit Court of Appeals and the limits are not currently in force.

19. National Institute of Occupational Safety and Health.NIOSH Pocket Guide to Chemical Hazards. DHHS (NIOSH) Publication No. 97–140. Washington, D.C. Government Printing Office, 1997. 284.

20. Cullen, Lisa. *A Job To Die For,* Common Courage Press, Monroe, Maine, 2002.p.75.

21. Cooper G.S., Germolec D., Heindel J., Selgrade M. Linking Environmental Agents and Autoimmune Diseases. Environ Health Perspect 1999 Oct: 107 (Suppl 5):659–660.

22. Hansen K.S., Isager H. Obstructive Lung Injury After Treating Wood with Sodium Hydroxide.J Soc Occup Med 1991 Spring; 41 (l):45–6.

23. Kardestuncer, T. and Frumkin, H.: Systemic lupus erythematosus in relation to environmental pollution: an investigation in an African-American community in North Georgia. Archives of Environmental Health, 1997, 52 (2): 85–90.

24. Florida Statutes, 440.15(5)(b).

25. 29 CFR 1904.5(a).

26. Letter Florida Division of Vocational Rehabilitation to Dale Goldstein, dated April 28, 1999.

27. Letter Social Security Administration to Dale Goldstein, June 18, 1999.

28. Summary and Opinion and Report of Autopsy, Dale Goldstein, 99–04801, Hillsborough County, Florida Medical Examiner Department, dated 11/25/ 99 and 9/23/99 respectively.

29. ibid.

30. Tyragiel, Josh. Holding the Checkbook. Time Magazine, September 9, 2002. p.65.

ADDENDUM CHAPTER 1:
NEWTOWN, GAINESVILLE, GEORGIA

Did Dale Goldstein, a physically fit and healthy 47-year-old man, contract Systemic Lupus Erythematosus (SLE) solely due to his extreme exposure to sodium hydroxide (NaOH) or lye, by inhalation and contact; or did he already have a genetic predisposition to SLE and the severe NaOH exposure exacerbated an already present, but dormant, condition?

No one else in the family appears to have SLE, and family members who have been tested are negative for the disease.

But what evidence is there of environmental causation in the onset of SLE?

According to researcher, Sullivan (1999), "If genetics were the sole basis of the disease, both identical twins either would always have the disease or would not have the disease." The actual risk for first-degree relatives of people with SLE to themselves develop SLE is only about three percent. Sullivan posits that the "genetic susceptibility is due to multiple genes, and that a certain threshold of genetic susceptibility must be reached before an external process is capable of triggering the disease."[1]

"Certain drugs have been shown to be linked to lupus, and it is possible that environmental toxins could induce lupus," according to the authors of the Agency for Toxic Substances and Disease Registry's (ATSDR) Public Health Assessment in the Gainesville, GA community."[2]

Researchers at the Lineberger Comprehensive Cancer Center, University of North Carolina were able to induce a lupus-like syndrome in pristine mice. "All inbred mice examined appear to be susceptible to this ... chemically induced lupus." The authors note that, "pristine-induced lupus in mice may provide insight into the causes of spontaneous (idiopathic) lupus and also may lead to information concerning possible risks associated with ... exposure to hydrocarbons in the environment."[3]

Gainesville, Georgia is a small town about sixty miles north of Atlanta. The town is often referred to as the "chicken capital of the world"[4] for the presence of its large poultry slaughtering and processing and animal feed industries.[5]

Other industries are also present in Gainesville as well; and are possible contributors to the air, soil, and water contaminants.[6]

Right in the industrial southern part of Gainesville is the predominantly African-American community of Newtown, consisting of nine streets, about 140 homes, and 300 long-time residents.[7]

The prevalence of Systemic Lupus Erythmatosis (SLE) in Newtown is "3 cases/300 persons (1000/100,000)", or a 6-fold increase on the highest reported prevalence to date. The incidence of lupus cases in Newtown is a 9-fold increase on the highest reported incidence: 3 cases per 4709 person years or 63.7 cases/ 100,000 years. The authors conclude "The results suggest that long-standing exposure to industrial emissions may be associated with an increased risk of lupus."[8]

Eighty-one households were surveyed, subsuming 246 residents.[9]

Excluded from the study were an additional twelve (12) unconfirmed cases in residents who had lived in Newtown in the 1960s and 1970s.[10]

ATSDR's Public Health Assessment in Gainesville could not go back prior to 1986 because emissions prior to that year were not required to be reported to EPA. Thus "it is impossible to determine to what extent area emissions impacted residential health prior to 1986."[11]

There is a possible other link between Dale Goldstein's death from SLE and the extraordinarily high incidence and prevalence of SLE in Newtown, Gaines-ville, Georgia: And that is the possible long term sodium hydroxide (NaOH) exposure in the so-called "chicken capital of the world."

Gainesville, and the Newtown area, are the sites of no less than seven major poultry slaughtering, poultry processing or poultry and animal feed preparation plants.[12]

Sodium hydroxide, the same chemical that led to Dale Goldstein's death, has several uses within the feed processing and poultry slaughtering and processing industries.

Caustic cleaners such as sodium hydroxide(NaOH), are used in meat and poultry processing plants for disinfection of facilities, disinfection of transport carriers, or to remove feathers from poultry carcasses.[13]

Sodium hydroxide (NaOH), may be used to decrease pathogenic bacteria, such as E. coli or salmonella in the "waste egg shell" which is used in feedstuffs, often for poultry.[14]

Sodium hydroxide (NaOH) has been studied in the pretreatment and fermen-tation of whole dead hens recycled into feed ingredients. "Feathers ... make pro-cessing of dead birds difficult"[15] "and degrade the quality of end-product meal because feathers have poorer digestibility." "NaOH or enzyme treatment ... may improve nutritional quality...."[16]

Sodium hydroxide (NaOH) is used in "scald and chill" procedures. The sodium hydroxide is found to have "no adverse effect on texture and color attributes of all (chicken) tissues" and that the NaOH extends shelf life of chilled broilers to 8 days "without adversely affecting quality."[17]

Sodium Hydroxide(NaOH) used as a food additive in animal drugs or feeds is generally assumed to be safe when used in accordance with "good manufacturing or feeding practice."[18]

The workers in the poultry and feed plants might have been exposed to varying amounts on a daily basis over a long period of time. Their family members, similar to the pesticide exposure of the children of farm workers and children who lived in the vicinity of farms where pesticides are being used, might also have been exposed to this, as well as the other identified toxins.

Finally, while different states regulate the waste products that result from meat and poultry treatment and byproducts differently, the sludge from wastewater treatment at poultry slaughtering and processing facilities has often been applied to farmland.[19]

We do not know to what extent sodium hydroxide was actually used in the poultry and feed industry processes in Newtown/Gainesville, its concentration if used, nor over what period of time.

Thus, while we cannot link the high incidence and prevalence of Systemic Lupus Erythematosus (SLE) in Newtown, Gainesville, Georgia to the extreme sodium hydroxide exposure that induced SLE in Dale Goldstein, there appears to be clear evidence of environmental causation of SLE.

References

1. Sullivan, K. The Complex Genetic Basis of Systemic Lupus Erythematosus. Lupus Foundation of America Lupus News, Vol. 19, No. 4, Fall, 1999, Vol 20, No. 1, Winter 1999–2000, p.l.

2. Petitioned Public Health Assessment, Newtown Community, Gainesville, Hall County, GA, Agency for Toxic Substances and Disease Registry, US Center for Disease Control, March 15, 2001,p.13.

3. Shaheen,V.M.;Sato,M.;Richards,H.B.; Yoshida, H.;Shaw, M.; Jennette, J.C.; Reeves, W.H. Immunopathogenesis of Environmentally Induced Lupus in Mice. Environ Health Perspect 1999 October; 107 (Suppl 5) : 723–727.

4. Gainesville, Georgia, Epodunk, www.epodunk.com/cgi-bin/genlnfo.php?loclndex =7957.

5. Petitioned Public Health Assessment, Newtown Community, Gainesville, Hall County, GA, Agency for Toxic Substances and Disease Registry, US Centers for Disease Control, November 19, 2002, Appendix C, Table 2, Toxic Release Inventory Sites, 1987–1998.

6. Petitioned Public Health Assessment, Newtown Community, Gainesville, Hall County, GA, Agency for Toxic Substances and Disease Registry, US Centers for Disease Control, March 15, 2001,p.2.

7. Georgia Department of Human Resources(GDHR). Newtown Neighborhood Cancer Investigation, Gainesville, GA. Atlanta, GA: GDHR; June7, 1990, as cited in Kardestuncer, T. and Frumpkin, H., 1997), Systemic lupus erythematosus in relation to environmental pollution: an investigation in an African-American community in North Georgia. Archives of Environmental Health, 52 (2): 85–90.

8. Kardestuncer, T. and Frumpkin, H.: Systemic lupus erythematosus in relation to environmental pollution: an investigation in an African-American community in North Georgia. Archives of Environmental Health, 1997, 52 (2): 85–90.

9. Petitioned Public Health Assessment, Newtown Community, Gainesville, Hall County, GA, Agency for Toxic Substances and Disease Registry, US Centers for Disease Control, March 15,2001, p.13

10. ibid.

11. Petitioned Public Health Assessment, Newtown Community, Gainesville, Hall County, GA, Agency for Toxic Substances and Disease Registry, US Centers for Disease Control, November 19, 2002.

12. ibid, Appendix C, Table 2, Toxic Release Inventory Sites, 1987–1998.

13. Vulcan Chemicals Technical & Environmental Services, Technical Data Sheet, 9/96.

14. Pacific NW Pollution Prevention Resource Center Pollution Prevention Research Projects Database, 9/98, at www.pprc.org/pprc/rpd/fedfund/usda/csrees/develop3.html.

15. Harvey, 1992, quoted in Kim, W.K. and Patterson, P.H., Recycling Dead Hens by Enzyme or Sodium Hydroxide Pretreatment and Fermentation, Poultry Science, 2000, 79:879–885.

16. Kim, W.K. and Patterson, P.H., Recycling Dead Hens by Enzyme or Sodium Hydroxide Pretreatment and Fermentation, Poultry Science, 2000, 79:879–885, p.884

17. Hinds, Margaret, "Effect of High PH Antimicrobial Dips on Physical, chemical and sensory quality of poultry", North Carolina A & T State University, Greensboro, North Carolina, cited on www.ag.ncat.edu/research/ projects/2000/effect_of_high_ ph__animicrobial__d.htm.

18. 21 CFR 582.1763 (4/1/97)

19. Meat and Poultry Processing Industry Sector Performance Program, US EPA, Meat Processor Wastewater Treatment Environmental Compliance Checklist, p. 4.

▼

CHAPTER 2

PUBLIX:
WHERE SHOPPING IS A
PLEASURE, BUT
DISTRIBUTION CAN BE
DEADLY

Publix Super Markets, Inc. is a fast growing southern company with 759 super-markets in Florida, Alabama, Tennessee, Georgia and South Carolina in 2003. The company posted sales of $15.9 billion for 2002. In Florida, Publix was said to be the state's largest private employer in 2003.[1]

When one walks into a Publix Super Market, the clean, welcoming, friendly environment is palpable. Personnel are pleasant and helpful. There is an amazing array of grocery choices. Freshly prepared sushi is being made as you watch. Publix truly lives up to its motto: "Where shopping is a pleasure."

But there is another side to Publix, and that is its distribution centers or ware-houses. Publix Super Markets Distribution Centers in Boynton Beach, Deerfield Beach and Miami, Florida reportedly made OSHA' s list of workplaces that had worker injury and illness rates of six or more job-related injuries or illnesses per

100 workers resulting in lost work time in 2001; in other words, double or more than double the average injury or illness rate.[2]

The reality is more horrifying. Between January 31, 1998 and July 11, 1999, three workers died from the injuries they received at the Deerfield Beach Publix Distribution Center. Two of those deaths occurred in the same way on the same type of equipment, within 10 months of each other.

And, these were not the first deaths to occur at that distribution center. In all, between 1988 and 1999, five workers lost their lives at Publix Deerfield Beach Distribution Center.

In 1988, the year the center opened, John Dunne[3] reportedly "got his foot caught on a pallet he was feeding into a pallet-loading machine. He was dragged off his forklift into the machine and badly mangled...."[4] John Dunne died from his injuries the next day. He was 38 years old.[5]

OSHA reportedly fined Publix a mere $350 and "ordered the company to install safety guards and an extra stop switch on the pallet machine."[6]

On August 3, 1992, Kenneth Smith was repairing light fixtures when he fell off a platform, plunging several feet onto a concrete floor and striking his head. Smith never recovered. He died five months later, on January 15, 1993, from his injuries.[7] Publix was fined $1000 by OSHA for "failure to install a guardrail."[8]

The 34-year-old Mr. Smith had financially assisted his aging parents on his income as a mechanic for Publix, where he had worked for seven years. Not only had they lost their son, and were left without his financial assistance, but up to one and one half years after his death, his parents were still receiving his outstanding medical bills, which should have been paid by Publix workers' compensation insurer. It took almost four and one half years before there was settlement with the grieving parents for a miniscule sum.[9]

Harris Jones, age 42, had worked for Publix for over 26 years. On January 31, 1998, Mr. Jones was "transferring product from a trailer to a large warehouse size freezer area. A hydraulic ramp (hinged to the warehouse) was utilized as the pathway between the truck and the warehouse." ... Because Jones was "unloading the last couple of pallets, his co-workers left him alone and moved on to unload another trailer. When he did not show up to help them, a co-worker went looking for him and found him pinned in an upright position between the ramp and the trailer door." His co-workers administered CPR, but it was too late.[10] He was pronounced dead at the North Broward Medical Center.[11]

"The most probable explanation", the OSHA inspector posits, is that after Jones unloaded the pallets, "he raised the ramp and turned his back to lock the rear doors of the truck. When he turned around, the ramp (which was raised but not to the locked position) was on its way down and pinned him against the truck door"…. Jones had "unknowingly bumped the down button as he walked by causing the ramp to lower itself onto him. The control buttons on the panel extended past the face of the control panel at the time of the accident." The inspector adds, "They have since been modified so that they are flush with the control panel."[12]

The OSHA inspection took place on February 2, 1998, three days after Harris Jones' death. It was begun at 11:00 am and in less than two hours the investigation and interviews were completed, and the closing conference had begun. No citations were issued. By the time the report was written, the protruding buttons had been fixed, according to the OSHA inspector. Despite the fact that the protruding buttons had been demonstrated to be a deadly danger, it was apparently considered a safety problem that had been resolved.[13]

It appears there were no OSHA citations, therefore there were no fines. This was the third death at this Publix Distribution Center. And as we shall see, protruding buttons on control panels would continue to be a deadly problem.

In Florida, the maximum workers' compensation payment is capped at a $100,000 death benefit and can only be paid if there is a dependent who is related, e.g., a spouse and/or underage children. The amount is based on a percentage of average weekly wages up to a ceiling, and is then prorated and paid out weekly until the maximum agreed upon benefit is reached. An additional amount, up to $5000, can be paid for funeral expenses.[14]

For the death of Harris Jones, the workers' compensation insurer apportioned the meager payments to the family of Harris Jones weekly over several years, after which time the ceiling had been reached and payments ended.[15]

Leonard Gale started working for Publix as a temporary employee in April 1998. In July 1998, he was hired as a permanent employee, unloading trailers.

On September 24, 1998, at about 9:00 am, Gale was unloading a trailer that had just returned to the warehouse. He was operating an electric tractor, sometimes called a motorized tugger. He apparently was backing into the tractor-trailer truck to unload four wooden pallets. In so doing, Leonard Gale backed into a metal load bar located about five feet from the floor and attached to the truck, used to keep the cargo in place. The bar struck Gale in the back and "pinned him to the control panel of the tugger." He was "pinned against the reverse button and the horn" and unable to move. The horn on the tugger was being pressed, which served to alert a co-worker, who ran into the trailer to try to free Leonard. "Smoke was emitting from the tugger when he went in, caused by the inability of the tugger to move."[16]

Paramedics responded to the scene and transported Leonard Gale to a nearby hospital, where he was pronounced dead at 9:44 am. The cause of death was "asphyxia as a result of chest compression."[17] Leonard Gale's young life was over at 28 years of age.[18] His widow was notified. Leonard Gale and his wife had no children.

A Publix representative interviewed by OSHA stated that the bar should have been removed before entering the trailer with the tugger.[19] Could it have been too dark in the trailer and he didn't see the load bar still up? OSHA's inspector thought there was enough light. But that opinion appears to be contested by other employees. The load bar in question was six to eight feet from the far end of the trailer.[20] The trailer was 48 feet in length.[21] And as we shall see later, even if he saw it, he still might not have been able to stop in time.

Apparently delayed by a storm, the OSHA inspector returned to the Publix warehouse on October 7, 1998, 13 days after this work fatality, to interview employees, one of whom asked that the Publix safety director be present during the interview. A Publix safety official then interrupted and said that the Publix attorney was on the phone and wanted to speak with the OSHA inspector. The attorney told OSHA's inspector that he was representing Publix and that the employee had a right to have the employer present. The attorney also stated that they (Publix) had an agreement with OSHA on this. It was later affirmed that no such agreement existed. But the pressure was on.[22]

OSHA' s conclusion: No violations, therefore no citations. There was not even a recommendation to fix the protruding reverse button so that this could not happen again.

Less than ten months later, on Sunday, July 11, 1999, at about 9:00 pm, Donald Boyd was unloading the remaining carts in the front end of a trailer in the cold storage section of the warehouse when his neck and back struck a load bar on the wall of the truck.

A co-worker was the first to arrive. He saw Donald Boyd, bleeding and unconscious, with the load bar pressing against his neck and back.[23] "... The tugger wheels were still in reverse and the battery plug on top of the tugger had to be pulled to stop the running tugger and free" Donald Boyd.[24]

"The impact of the collision appeared to have forced the victim's chest area against the 'control/console' of the tugger. This in turn compressed his chest cavity to the point where he was not able to inhale/breathe.... Fellow employees were alerted to the situation by the sounding of the tugger's horn, which had been activated by the victim's chest. They responded to the scene and proceeded to free him. Paramedics were subsequently called and the victim was transported to the hospital. He was pronounced dead a short time later."[25]

The OSHA inspection notes that the tugger "has one reverse speed and when the tugger starts off in reverse it quickly achieves this speed and can only be stopped in approximately a five foot space when the hand brakes are quickly applied. An operator inside a trailer when backing at this single speed who suddenly sees an obstruction, such as a load bar in this case, would have insufficient time and distance to stop the tugger safely...."[26]

This deadly safety flaw was also identified in the Investigative Action Report, issued by the Sheriff's Office, in which it was noted that a Publix management official is aware that "there is no automatic escape mechanism manufactured into the workings of the tugger which would free an operator during this type of event." The report concludes: "After investigating the scene of the accident and reviewing the findings of the medical examiner, I firmly believe this accidental death could have been prevented had the 'Tugger' been equipped with some type of 'escape' or 'failsafe' device."[27]

At age 25, Donald Boyd, loving father of a four-year old son, was dead. His young life was crushed out of him, despite the known danger of the protruding reverse button on the tuggers, after the death of Leonard Gale ten months earlier.

One possible contributing factor in the death, that never came up in the investigation, is the use of a "computerized efficiency system" that sets a productivity standard of so many loads per hour. If an employee misses their quota a number of times, it is reported that they can be disciplined, suffer pay cuts or be discharged. Workers blame the system for increased accidents and increased severity of injuries.[28]

Donald Boyd had started working at Publix less them one month prior to his death. It is possible that he was behind on his quota at the time he was pinned in the truck, and was working off the clock while his co-workers were on break. According to one professor of occupational medicine, "Any time you speed up the line, you're setting yourself up for potential accidents."[29]

The workers' compensation benefit was paid by a staffing company. Mr. Boyd had been hired through the staffing service and assigned to Publix.[30] Thus, the weekly death benefit payment would have been the responsibility of the staffing company, not Publix.

Ultimately, OSHA cited Publix Supermarkets for modifications affecting safe operation made without the manufacturer's prior approval, and markings not in place, or illegible, on the machine. Both were deemed serious violations. Publix was fined the grand total of $7000. A third violation, "operating at speeds that don't permit stopping in a safe manner," was deleted. One might justifiably ask how many workers must die before OSHA issues a meaningful violation and penalty.[31]

Perhaps part of the answer lies in a letter from OSHA's Area Director to a Publix Corporate Safety Specialist on November 4, 1999, in which he states:

"There is no protection from striking objects for tugger operators and the reverse button on newer tuggers is exposed to accidental contact and activation.

"Since no OSHA standard applies and it is not considered appropriate at this time to invoke Section 5(a)(1) of the OSH Act, the 'general duty clause,' no citation will be issued for these hazards.

"In the interest of workplace safety and health, however, I recommend that you take the following steps voluntarily to eliminate or reduce your employees' exposure to the hazards described above.

"Retrofit all tuggers with a cage system which would prevent objects from striking the operator from all sides and above. The cage system would be similar to what is used on forklifts. Retrofit all tuggers having an unprotected reverse switch with a button guard. The button guards are similar to the ones used on machinery where accidental activation must be avoided and is basically a metal frame that fits under the button and comes out to act as a recess for the button.

"Retrofit all tuggers so that both the reverse button and the speed lever must be activated in order to back the tugger up."[32]

The Section 5(a)(1), or the "general duty clause" referenced in the second paragraph of the letter, refers to a section of the Occupational Safety and Health Act (OSH Act), which states that "each employer shall furnish to each of his employees employment and a place of employment which are free from recognized hazards that are causing or are likely to cause death or serious physical harm to his employees." The general duty clause is used when there is no specific standard that applies to a particular hazard.[33] One has to wonder why the general duty clause was not invoked in this, the fifth fatality, at the facility.

The criteria set forth to meet the general duty clause standard are:

a.   "The employer failed to keep the workplace free of a hazard to which employees of that employer were exposed.

b.   The hazard was recognized.

c.   The hazard was causing or likely to cause death or serious physical harm; and

d.   There was a feasible and useful method to correct the hazard."[34]

This criteria would indeed appear to have been met by Publix. But OSHA chose not to invoke Section 5(a)(1). Instead it recommended the changes that could save lives.

And in recognition of the lack of OSHA teeth, the letter from Publix to OSHA, in response to OSHA's citation, notes that Publix now has the "proper operator instruction decals" and "legible horn button markings and reverse control switch button markings." OSHA's retrofit recommendations that would save lives were not mentioned.[35]

Worker lives have been lost. Families are left with their grief, their pain, their loss, and their financial loss. Publix finally received a slap on the wrist, in the form of a $7000. penalty.[36]

By comparison, CBS is facing a fine of $550,000 for airing Janet Jackson's Super Bowl performance in which one breast became exposed.[37]

We truly need to examine our priorities, when exposing a part of anatomy on TV is valued at almost eighty times the loss of a worker's life.

References:

1. Tamen, Joan Fleischer. Publix Workers Vote on Union. South Florida Sun-Sentinel, June 5, 2003 p. 1D.

2. Tamen, Joan Fleischer. Injury, illness alert issued. South Florida Sun-Sentinel, March 11, 2003, p. 1D.

3. All names of the deceased in this chapter have been changed to protect their privacy and that of their families.

4. Meyer, Harris. Publix, where working can be lethal. New Times Broward-Palm Beach Feature, February 10, 2000, www.newtimesbpb.com/issues/2000–02–10/feature.html/4/index.html

5. State of Florida, Department of Health, Office of Vital Statistics. Certificate of Death, Certificate Number 88 080835, dated 8/28/88.

6. Meyer, Harris. Publix, where working can be lethal. New Times Broward-Palm Beach Feature, February 10, 2000, www.newtimesbpb.com/issues/2000–02–10/feature.html/4/index.html

7. State of Florida Department of Labor and Employment Security, Division of Workers Compensation, Petition for Benefits re _____.

8.   OSHA Inspection No. 109681874, Violation Summary; and Meyer, Harris. Publix, where working can be lethal. New Times Broward-Palm Beach Feature, February 10, 2000, www.newtimesbpb.com/issues/2000–02- 10/feature.html/4/index.html

9.   State of Florida, Department of Labor and Employment Security, Office of Judge of Compensation Claims, re _____.

10. OSHA Inspection # 300496882, Safety Narrative.

11. State of Florida, Department of Health, Office of Vital Statistics, Certificate of Death, Certificate No. 98 011875.

12. OSHA Inspection # 300496882, Safety Narrative.

13. ibid.

14. Florida Statutes, Chapter 440.16.

15. State of Florida Department of Labor and Employment Security, Division of Workers' Compensation, Claim Cost Report for _____, dated February 11, 2003.

16. Sheriff's Office, Broward County, Florida, Investigative Action Report, Case # DR98–09–03615.

17. ibid.

18. State of Florida, Department of Health, Office of Vital Statistics, Certificate of Death, Certificate No. 98 115040, dated 9/24/98.

19. OSHA Inspection Number 300499936, Opening Conference Notes.

20. Sheriff's Office, Broward County Florida, Investigative Action Report, Case # DR98–09–03615.

21. OSHA Inspection Number 300499936, Opening Conference Notes, p.3.

22. ibid, p.4.

23. Sheriff's Office, Broward County, Florida, Investigative Action Report, Case # DR9907–1739.

24. OSHA Inspection Report, #301886644, Opening Conference Notes.

25. Sheriff's Office, Broward County, Florida, Investigative Action Report, Case #DR9907–1739.

26. OSHA Inspection Report, #301886644, Citation 01, Item No.001(c).

27. Sheriff's Office, Broward County, Florida, Investigative Action Report, Case #DR9907–1739.

28. Meyer, Harris. Publix, where working can be lethal. New Times Broward-Palm Beach Feature, February 10, 2000. www.newtimesbpb.com/issues/2000–02–10/feature.html/4/index.html.

29. ibid.

30. State of Florida Department of Labor and Employment Security, Division of Workers' Compensation, Office of Judge of Compensation Claims re _____.

31. OSHA Inspection Report, #301886644, Citation and Notification of Penalty, November 10, 1999.

32. Letter Area Director, US Department of Labor, OSHA, Ft. Lauderdale Office to Publix Super Markets, Inc., Deerfield Beach, Florida, dated November 4, 1999, in OSHA Inspection Report, #301886644.

33. 29 CFR 1910.5 (f) as cited in OSHA Field Inspection Reference Manual, CPL 2.103, Chapter III. Inspection Documentation, p.8.

34. ibid.

35. Letter Corporate Safety Department, Publix Super Markets, Inc. to OSHA Area Director, Fort Lauderdale Office, dated December 30, 1999, in OSHA Inspection Report, #301886644.

36. OSHA Inspection Report, #301886644, Inspection Detail.

37. Wire Reports, Washington D.C., South Florida Sun-Sentinel, July 1, 2004.

CHAPTER 3

▼

THE GROWING FIELDS, THE KILLING FIELDS

Farm workers work from dawn to dusk to put food on your table and mine that they can barely afford for their own. They are indeed the least powerful, and the poorest, workers in our nation. Farm workers are excluded from minimum wage laws and other labor protections. Yet they are working in one of the most dangerous occupations in our country. Fatality rates for farm workers in 2002 were 22.7 fatalities per 100,000 workers, second only to mining deaths of 23.5/100,000.[1]

Tirso Moreno came to the US from Mexico in the early 1970s. Mr. Moreno was himself a farm worker for over 20 years. He remembers Cesar Chavez, the early days of fighting for farm worker rights and decent working conditions. Now Tirso Moreno is the General Coordinator for The Farmworker Association of Florida, Inc., headquartered in Apopka, Florida, deep in the growing country of central Florida.

Legal age in most industries is 18 years of age. In farm work, at 16 years of age one is considered an adult. There is lots of child labor below age 16 years in the industry too, explains Moreno, even though that is illegal. In a 12-year study at a Wisconsin trauma center, fully 22% of patients sustaining farm injuries were under 16 years of age.[2]

In order to understand the magnitude of the worker injuries in farm work, one has to understand how the work is done.

The grower hires a labor contractor; who in turn hires, pays, transports, and supervises at the work site. Labor contractors compete with each other: the lowest bid is likely to get the job, explains Moreno.

In Lake Placid, Florida in 2002, some 700 farm workers reportedly were held in involuntary servitude; while labor contractors "deducted debts from wages, and used violence and threats to restrict workers from leaving." The contractors were convicted, "but the growers who used this indentured labor paid no penalty—and the workers were never compensated for back wages."[3]

In Central Valley, California, in September 2003, after picking grapes and tomatoes under the hot sun for weeks and not getting paid, 48 farm workers decided to take their fight for their combined $87,480 in unpaid wages to the legal system; and sought assistance from the California Rural Legal Assistance Foundation, which sought restitution from the bonding company. Under California law, farm labor contractors are required to post a bond based on payroll amounts.[4]

In some strawberry fields in California, a machine called a "pace setter" is used to increase productivity. The machine's conveyor belts, across 15 rows of strawberries, move through the fields at half mile per hour. Workers bend, filling boxes, while selecting only acceptable quality of the fragile berries. The machine's speed necessitates that while a worker is bending from the waist, berries be picked with both hands and then lifted in twenty-pound boxes, two at a time, onto the conveyor belt. Sometimes the work starts at 6:30 am and doesn't end until 5 pm. "We could not drink water nor go to the bathroom, because they are too far and we lose pace with our co-workers," stated one worker.[5]

For Antonio Valdez,[6] that would prove deadly. It was over 100 degrees on July 28, 2004, in the Central Valley of California, when Antonio Valdez was picking grapes. The 53-year-old Valdez was strong and healthy, as confirmed later in the coroner's report. But he was under pressure from the foreman to meet his daily quota of 50 crates. So he didn't take time to stop for water. At the end of his 10-hour shift, Antonio Valdez collapsed and lost consciousness. A 911 call was made but when he came to, the foreman cancelled the ambulance. His son proceeded to drive him home in the hot, non-air-conditioned car. Antonio began vomiting and went limp. His son drove to the nearest hospital, but it was too late. Within minutes, Antonio Valdez was dead from heat stroke. He had been harassed and made afraid to stop for water in 100-degree heat. Immediate medical attention was denied. The company claimed proper procedures were followed.[7]

Less than one year later, in July 2005, Salvadore Rae died in the 105-degree heat of the bell pepper fields of California.[8] The same month, Angel Greg was found dead in the fields the morning after picking grapes for 10 hours; and two other farm workers collapsed, one while picking melons, the other in tomato fields.[9] By the time 24-year-old Carmine Cal collapsed in the tomato fields, after working nine hours in 100-degree heat, with only one break and a 20-minute lunch period; he was the fifth farm worker death from heat in the California's Central Valley in 2005.[10]

Farm workers are paid by the piece, and that, says Tirso Moreno, is what creates the danger.

The orange industry in Florida requires upwards of 100,000 workers. "An average orange worker will get paid $.65 for a 90-pound box of oranges," Mr. Moreno tells me in our interview in September 2002. That price includes setting up the ladder, climbing into the tree with a sack, reaching out from the ladder to pick the fruit, putting the fruit into the sack, which is getting heavier as the worker goes along, coming down the ladder with the filled sack, walking to the worker's individual container, which is often at some distance, putting the contents into the container, pushing the ladder from tree to tree, and finally pushing the filled container to the truck." And, "if you leave fruit on the tree or on the ground, you have to go back and clean it up." All this incredibly back-breaking labor, under horrendous working conditions, earns a farm worker 65 cents.

Tomatoes are the second largest crop in Florida. The pay is $.40-$.50 for a 32-pound bucket at the time of this interview. According to the Coalition of Immokalee Workers, that rate hadn't changed much in twenty years.[11]

Unlike oranges, there are no ladders, no climbing trees, but bending down, and lots of it. Here one has to choose the right tomatoes, no pink ones, no stems, and have to lift the bucket up onto the truck, sometimes after running to catch up with it, explains Moreno.

The low wages contribute to make the job dangerous. "And the rates have remained the same, despite rises in cost of living. Even $.01/lb. more would make a difference, or pay by the hour," suggests Moreno. Farm workers don't even earn minimum wage: The average wage is still only $7500/year.[12]

While wages for farm workers remain extraordinarily low, agriculture is big business. In Florida, it is the second largest industry in the state, with an estimated $60 billion plus per year economic impact.[13]

There are no benefits, no medical coverage for most farm workers. According to Moreno, one only goes to a physician when it's an emergency, and the same is true for the families, the children. In thirteen states, there is no coverage for occupational injuries or illnesses in agricultural workers. Other states may have exclusionary restrictions.[14]

Most farm workers are financially eligible for food stamps, and other governmental needs-based programs. Yet, only 17% of farm workers have used such programs.[15]

Because there's no improvement in one's life over the years, if other work is available one takes it, explains Moreno, causing high mobility in the field, and a lack of knowledge of the dangers of farm work, this most dangerous industry.

The most insidious of all the dangers in farm work is the exposure to pesticides, which can cause permanent disability and death.

The law requires initial training with pesticides within five days. But the agencies don't enforce that, Tirso Moreno explains. Farm workers can thus end up performing with pesticides, before the training. Enforcement is often left to the local level, resulting in wide variations in compliance, e.g. California's 57% compliance with training, North Carolina's 35% compliance.[16]

Until 1995, workers didn't even have the right to know to what they were being exposed. Now there is a workers protection standard, explains Mr. Moreno, referring to the EPA regulation governing pesticide safety training, notification, protective equipment, decontamination supplies, and emergency medical assistance, as well as OSHA's Field Sanitation Standard. But rules are not enough. The industry fights very hard to protect its interests. "Growers want to harvest when the fruit is ripe, and they don't care about (safety). There is no complete safety with pesticides," states Moreno, "there is always a residue."

Environmental Protection Agency has established "re-entry periods," depending on the toxicity of the chemical used. A re-entry period is a time before which a worker cannot re-enter the field. Workers are supposed to be notified, thru posting, as to the location where the employer has applied what chemical within the last 30 days, and identify the re-entry period.

Arcury, et al. (2001) found that only 48% of workers reported that employers had informed them of pesticide application or posted signs.[17]

Tirso Moreno cited a massive incident of chemical poisoning of farm workers in Hillsborough County, Florida. In this instance, the consequences were immediate. As a rule, the cancers that result from the chemical poisoning don't show up for years.

A study of cancer incidence among farm workers in California, in comparison to all Hispanic populations in California, shows elevated incidence for brain cancer, and skin melanoma; and significantly elevated incidence for stomach cancer, cancer of uterine cervix, and uterine corpus, and leukemia. It was also noted that a lower percentage of farm workers were diagnosed at an early stage of their disease.[18]

It is impossible to look at work injuries and deaths among farm workers, without also examining the appalling conditions in which most farm workers are forced to live.

Typical housing for a farm worker might be a very run down mobile home that is rented at as much as $1000/month, so groups of 10 or more live together, just to make ends meet. A worker in that situation is unlikely to have access to a washing machine to get rid of pesticide residue on one's clothing. The inability to effectively wash skin or work clothing contributes to acute pesticide poisoning.[19]

The Unitarian Universalist Service Committee's Just Works Program reported similarly inhumane conditions at a migrant farm worker community in Crewport, Washington, described as "no clean water, no bathrooms, no garbage removal, no medical care." The only water was the Columbia River, itself polluted with runoff pesticides. Farm workers are said to have had to go door to door to beg for drinking water.[20]

Steve Meyers is an attorney in Immokalee, Florida, the heart of farm country. He estimates he sees about 80–120 cases a year. I asked Steve what happens to Florida's farm workers once they are injured.

Because of the nature of the work, Meyers explains, typically he sees lower back injuries from lifting, injuries from falling out of trees, rattlesnake bites, and exposure to chemicals.

The chemical exposure is difficult to diagnose because so many symptoms mimic other diseases. There are not many toxicologists, or neurologists, and fewer still willing to work for workers' compensation, Meyers adds.

The problem is compounded by the poor language skills of the injured worker, and that the worker may not know the chemical to which he/she was exposed. In a disputed claim, the burden of proof falls on the injured worker. So farm workers, with limited language, often poor historians, are placed in the position where they have to prove their medical diagnosis, what they were exposed to, and that the exposure caused the disease. The "what" to which exposed can be determined, of course, but it's a time-consuming process, says Meyers. "And the farmer often threatens to call immigration if the worker is an undocumented worker."

Generally, the farm worker is sent to a walk-in clinic with which the grower has an established relationship. The clinic doesn't order a full range of blood work, doesn't refer to a toxicologist. They tell the worker nothing is wrong, go back to work.

If it's an obvious injury, e.g. a broken bone or herniated disc, they'll get treatment, even if delayed. The obvious stuff gets treated, says Meyers, but they are quick to release back to work. When the worker returns, the grower tells him don't bother coming back, unless you can do full duty. Large farms have a steady supply of workers. Generally, the injured worker is fired after a few weeks.

In Florida, once the injured worker has reached maximum medical improvement (MMI), the physician gives an impairment rating. A lower back injury, non-surgical, might be a 3–5% rating, explains Meyers. What this means in real life terms is that after MMI, the impairment rating is multiplied by 3 to determine the number of weeks of checks at one-third of wages. He gave the example of a 5% work impairment x 3 = 15 weeks of checks at 1/3 of wages. This, then, would be the final compensation for lost wages. The medical case can be kept open. But what does this person do for a livelihood thereafter?

A typical settlement for a small permanent injury might be $5000 after most treatment. MMI effectively ends the compensation checks.

With a seriously injured older (over 50) worker with a permanent injury, the law does allow to file for lifetime permanent disability benefits. Then workers' compensation will try to make a settlement. If that person doesn't have legal documents (undocumented immigrant), the insurer can say he's not allowed to work legally. The employer sends a letter to the worker saying he will let the worker work if he has valid ID. This wasn't a concern until the worker got injured, notes Meyers.

What about the farm workers who die on the job? In Florida, Meyers explains, survivor workers' compensation benefits are capped at $100,000. However, that amount is distributed to the spouse and underage children at the 2/3 wage rate set by workers' compensation, until children reach maturity, until spouse remarries, and in no case beyond the $100,000 ceiling. What it is not, is a lump sum settlement of $100,000. Nationally, compensation for farm worker deaths is substantially below other industries.[21]

An employer can't be held liable for negligence. "There is an immunity unless you can prove they did something so bad it was intentional, or substantially certain to cause injury, for example, removing all the safety features on a machine and a worker gets his arm cut off. Even in such an instance the courts have made it very difficult," Steve Meyers notes.

One is dealing with an unsympathetic employer, a clinic or physician being pressured by the insurer, and the defense attorney is being paid to save the insurer money.

"We have given up our negligence remedies (referring to the quid pro quo of workers' compensation) and yet the benefits keep being whittled down," states Steve Meyers.

*   *   *   *

Moving slowly down the main street in Salinas, California on a warm July day in 2004, was a strange constellation of what appeared to be two outhouses on wheels trailing behind an ancient, decrepit bus. The former are the primitive bathroom facilities provided to the farm workers in the fields. Often such facilities can be a mile or more away, and lacking either water or paper.

Salinas is in the heart of growing country in California. It is here that I met Jorge Fernandez, 46, and Guillermo Ruiz, 45 years of age. Both men have been exposed to pesticides, over many years, in their work in the fields of California.

"I worked with methyl bromide for eleven years," Jorge Fernandez explains. He sighs deeply. "They never told me about the effect that it has. They just told me it smelled like chlorine, and that's what made me cough, my eyes water and run. They never gave me equipment, no masks, no protection. Now that I am sick, I see that everything they were doing was wrong." Guillermo Ruiz, also a victim of severe and sustained pesticide exposure, quietly affirms his colleague's description.

The chlorine-like smell, that Mr. Fernandez describes, is likely chloropicrin. When methyl bromide is used as a fumigant, it is often mixed with chloropicrin, intended as a warning agent, because of its intense odor.[22]

Fernandez explains how the methyl bromide is used: "The company would apply methyl bromide and then cover it with plastic. They had to leave it on for 24 hours, before we could go in there to work. But they never respected the 24-hour requirement and would make us go in immediately. We would take off the plastic and work, and sometimes we would see dead animals on top of the plastic, such as deer, dogs, cat, birds, and we would just move them aside because we had to keep on working. The deer, we would have to drag with a tractor to the side of the road, so that we could work. We would go in 4 or 5 minutes after they had cut the plastic. They would only cut one aisle, not the other. There would be a second aisle that would be left without being cut."

Methyl bromide is used as a soil fumigant in strawberry fields across California, like the ones in which Jorge Fernandez and Guillermo Ruiz worked. The methyl bromide is injected into the soil to a depth of one or two feet before planting. The soil is then covered with plastic tarps, which are to be removed 24–

72 hours later in the case of strawberry growing, or at the end of the entire season, in the case of tomatoes.[23]

Georgina Mendoza, California Rural Legal Assistance, Inc. attorney and our translator, explains that in this case a separate company goes in and injects the methyl bromide, or any type of fumigant, and covers it with plastic. The company for which Fernandez and Ruiz worked was solely in charge of cutting and removing the plastic tarps. There is a legal time requirement between the time they cut the plastic and the time the plastic is removed. For that type of fumigation method, there has to be at least 24 hours between the time they cut the plastic and the time the plastic is removed, to therefore allow the material to dissipate, and minimize the effect on workers. In this case, what they were required to do for eleven years, was to cut the plastic and remove it immediately, or within five minutes, thereby being exposed to methyl bromide, and chloropicrin, and other chemicals in large doses. There are rows in the field. Legally, the company that cuts the tarps has to cut every aisle down the middle to assure that as much chemical as possible gets out. After 24 hours, after cutting every single aisle, then supposedly it is safe to remove. What would happen is that they would cut one aisle, not the other, alternating, and then remove the tarps. This was how Mr. Fernandez and Mr. Ruiz were exposed to unsafe quantities of methyl bromide. Many times, there would be a bubble that would form under the tarp that would burst, and the gas was really strong. The company would tell them to remove the plastic at night, so the inspectors would not see.

Methyl bromide is highly toxic. Methyl bromide exposure causes injury to brain, nerves, lungs, and throat. High doses also damage liver and kidneys. Contact with skin and eyes cause irritation and burns.[24] "It is most dangerous at the actual fumigation site itself.... [I]nitial symptoms include weakness, despondency, headache, visual disturbances, nausea, and vomiting. Later central nervous symptoms emerge, including numbness, defective muscular coordination, tremor, muscle spasms, lack of balance, extreme agitation, coma, and convulsions.... [G]ross permanent disabilities or death may result."[25] NIOSH ranks it as a potential occupational carcinogen.[26] Prolonged exposure at levels above 200 ppm (parts per million) results in death, usually from irreversible pulmonary edema.[27]

Methyl bromide is widely used for fumigation, particularly of strawberries and tomatoes. Every year, about 42 million pounds are used in agriculture in the United States. Strawberry and tomato crops alone consume about 14 million pounds of methyl bromide a year. "About 80 to 95% ... eventually enters the atmosphere."[28] For that reason, methyl bromide has been banned by the European Union.[29]

"It is almost a year now that I have not worked, that I have been disabled." "I can't breathe, and I forget a lot of things." Fernandez gets earaches, headaches, chest pains, and a tightness in his chest. His throat is also affected, and his vision. He describes pain in his skin, and vision "that feels burned." He also has depression. Sometimes his doctor doesn't have the medicine he needs, and he doesn't have the money to buy it, so he goes without.

Georgina Mendoza explains that the California Rural Legal Assistance, Inc. is doing community education on the nature of methyl bromide, and the use of protective clothing, but not a lot has changed.

"All the growers should know the dangers." The nature of the work is seasonal, thus there are many new workers. "It is not in the growers' interest to tell the workers (the health risks)." There is a legal requirement to provide safety training and protective gear, but the current laws are not being enforced, Georgina explains, and the responsibility for enforcement is put at the local county level.

Jorge Fernandez continues, "Here, (there are) a lot of schools nearby, the schools are right there and there are fields all around it.... [A] nd they use so much methyl bromide, they apply it next to schools. I, who have worked with it for about 11 years, have seen the kids are out there playing, while we are working with watery eyes. And I think to myself, how can it be possible that they allow those kids to breathe it? That isn't right, that affects all human beings."

In 2002, in California alone, 172 million pounds of pesticides were applied to the fields. The kind of chemical drift of which Jorge Fernandez speaks sickened a reported 478 people that year.[30] Many more go unreported because victims or physicians may not identify that the symptoms they are seeing are associated with pesticide exposure.[31]

"Children are more vulnerable ... and their ability to detoxify chemicals is limited. Their exposures to pesticides from all pathways (food, water, air, and contaminated surfaces) are likely to be higher, because they eat more food, drink more water, and breathe more air per pound of body weight. Exposures early in life can cause impaired growth and development, ... and lifelong disabilities."[32]

Ironically, in Duval County, Florida, the Environmental Protection Agency reportedly recruited parents of infants in poor neighborhoods to participate in a study of the effects of pesticides on infants. The study was said to be underwritten by the American Chemical Council. Parents were to be paid $970, a camcorder, bib and t-shirt to allow their babies to be exposed to pesticides during the two-year study. Fortunately, following adverse publicity, the study was cancelled.[33]

Jorge Fernandez continues, "I didn't know that I wasn't supposed to wash my clothes along with my children's clothes. The methyl bromide is so delicate, that you can't even wash your clothes together with your family's. I didn't know that. I hope to God, my children, they don't have this. My children don't have anything right now, thank God.... I need help so that I can get cured from whatever it is that I have. That way nothing occurs to them the way it did to me."

Fernandez' physician says his disability is permanent. The disability office says he is totally disabled, but temporarily.

The physician gives him pills for pain and for depression. "That's it. He just looks at me and (treats symptomatically). "I'd like to see a professional doctor that knows about this, to get a blood analysis, an MRI. I want to know the cause of why I can't breathe, and why I forget things, and why I have blurry vision, and watery eyes. When I get pain, like a squeezing in my chest, my doctor tells me to take a pill for an anxiety attack."

He says he tells his doctor: "I'm going to take you to a funeral and that guy (the deceased) looks better than me."

What lies ahead for Jorge Fernandez and Guillermo Ruiz? There is a general "lack of knowledge by health care providers on how to recognize and treat pesticide-related injuries and illnesses."[34] What both men really need is to be seen, diagnosed, and treated by specialists who understand toxicology and environmental medicine. Georgina Mendoza is trying to arrange for at least a diagnostic visit, on a pro bono basis, at a university medical center for the two men.

Their employer didn't have workers' compensation coverage, didn't have any insurance. Thus Fernandez and Ruiz don't have access to even the treatment provided under workers compensation. Both men are receiving state disability, but that will soon run out. So there is no money for the medical evaluation and treatment.

To get workers' compensation for pesticide exposure is difficult and complex. Causation is often difficult to prove, i.e., that the illness happened from the work.

Studies of farmer sprayers in the Netherlands demonstrated statistically significant (unlikely to have occurred by chance) sensory nerve and motor nerve conduction velocity reductions.[35]

Even when the causation is well established, and accepted by the workers' compensation court, still injured workers may have to continue the battle for benefits for years, as in the Henson case.

G. J. Henson worked as an agricultural mechanic for United States Sugar Corporation for 28 years, until he became too disabled to work. Henson repaired broken or malfunctioning equipment in the fields and was regularly exposed to pesticides, i.e. 2,4-D ametryn, atrazine, parathion, mocap, malathion, paraquat, azodrin, dursban/chlorpyrifos, guthion, diazinon, dalapon, methal arsenic acid, asulox, and polado. He, too, was told that the pesticides would not harm him, and was given no training on safety precautions, and only gloves that were unsuited for the work he needed to do, or latex gloves that tore easily. He suffered from shortness of breath, nausea, gastritis, and muscle weakness, dating back to 1977. He was treated in the company's medical clinic. Finally in 1996, he began seeing his own physician for weakness, dizzy spells, and shortness of breath. He was referred to a pulmonologist and diagnosed with a paralyzed phrenic nerve, which had resulted in a partial collapse of one lung, leaving him confined to a wheelchair, and dependent on a ventilator. Despite causation evidence and four expert witnesses, US Sugar appealed the case to Florida's Supreme Court, questioning whether Henson's expert witnesses met the Frye Test, i.e., established causation to meet the standard established in Frye v. United States, 293 F. 1013 (D.C.Cir.1923), considered a more difficult standard to meet. The Florida Supreme Court upheld Henson's causation, stating that it is "generally accepted in the scientific community that organophosphates are neurotoxic." But the court also upheld the Frye test.[36]

Thus US Sugar succeeded, not only in delaying settlement of this case, but in making it more difficult for workers in future Florida cases, exposed to other toxins, to prove their case. The Florida Fertilizer & Agrichemical Association, the American Chemistry Council, and growers are said to have filed briefs in support of installing the Frye test in workers' compensation court, as they saw it, to keep out "junk science."[37]

This is just the kind of tactic that succeeded in delaying, for more than sixty years, the removal of lead from gasoline, despite its well-documented devastating effect on children's brains.[38] Industry-financed disinformation campaigns kept tobacco firms out of the reach of regulatory agencies for years.[39]

Jorge Fernandez tells me he wants my book to go into school libraries, so what happened to him doesn't happen to them. "… If kids learn at an early age about the effect, the dangers, of methyl bromide, then they won't have the same problems that I do. Because they don't study this right now, they don't know. They're not going to be able to study when they are older. Because if I'm forgetting things, imagine when they're in high school. They won't be able to study when they are older," Jorge says, explaining he is referring to the effect of the methyl bromide on one's brain. "The way I feel, I don't want anyone else to feel," he adds.

"There are classes for math, there should be a class for chemistry that shows the chemical effects of pesticides, the destruction and its causes, what they do to our planet, the air, the water. Educate our children … teach the kids the symptoms. The principals, they could go outside and say 'no, don't put it (the methyl bromide) here. Put it on Bush's (referring to the US President) ranch.' There are lots of schools around here."

Ozone depletion is linked to skin cancers, cataracts, other damage. Many consider methyl bromide to be the most dangerous of all the ozone-depleting substances, and the one, which if eliminated, would have the greatest impact on reduction of ozone loss.[40]

"Now I have become aware of the ozone layer. How is it possible that people are allowing this to poison our planet? These are educated people and they are doing this to our planet!" Jorge states poignantly.

We need to "tell the higher ups what they are doing to us, that they are really messing up, and they need to stop it now. And those are people who are educated. If it's like that, I'd rather be the way I am. I can't see or write."

"They tell my kids to study and do good. And that's what I try to teach, because it will be better for them to do good than bad. Money and power, at the end of the road, isn't much. When you leave this planet, you should leave something good for everyone."

To date, none of the Fernandez or Ruiz children are working in the fields. "They are doing everything in their power to make sure their children get an education," Georgina Mendoza tells me.

"We are destroying the most important things that there are in life: Your self, your life, your planet. And that's what we destroy."

"If Bush put his daughters on a ranch and one of his daughters got sick(from the pesticides), there would be action. If we die, who cares? We've jumped over the border."

As I was leaving, Jorge Fernandez walked me to the door and showed me a picture on the wall of a cluster of small, worn, very crowded houses on a narrow dirt street. This is the California farm worker community where he grew up as a young child. His father, too, had worked in the fields of central California.

References

1. US Bureau of Labor Statistics, Census of Fatal Occupational Injuries, cited in *Death on the Job: The toll of Neglect*, AFL-CIO, 13th edition, April 2004, p. 26.

2. Cogbill, Thomas H., Steenlage, Eric S., Landercasper, Jeffrey, Strutt, Pamela. Death and Disability From Agricultural Injuries in Wisconsin: A 12-year Experience with 739 patients. Journal of Trauma, 1991; 31(12): 1632–1637.

3. Powers, Nancy R. Legislative Action could bring justice in the fields. South Florida Sun-Sentinel, April 28, 2003, p.l9A.

4. Furillo, Andy. The Central Valley laborers say they're owed $87,480. Sacramento Bee, December 24, 2003, reproduced in "'Farmworkers Fight to Get Paid," California Rural Legal Assistance Foundation at http://www.crlaf.org.

5. United Farm Workers of America, AFL-CIO, Newsletter, undated, on or about July, 2004.

6. All names of the deceased in this chapter have been changed to protect their privacy and that of their families.

7. Barbassa, Juliana. Report: Heat stroke, a concern every summer, killed farm worker. The Associated Press, in San Diego Union-Tribune, August 18, 2004.

8. After second farm worker dies from heat, UFW calls on governor and lawmakers for emergency action. United Farm Workers, 7/18/2005, www.ufw.org.

9. The grapes of wrath, again. The Economist, September 10, 2005, p. 36.

10. Martin, Mark. State seeks to halt farm worker deaths. New rules on breaks after 5 lose their lives in heat of summer. San Francisco Chronicle, August 3, 2005.

11. Coalition of Immokalee Workers Boycott Taco Bell Fact Sheet, undated.

12. National Agricultural Workers Survey data, cited in Powers, Nancy R. Legislative Action could bring justice in the fields, South Florida Sun-Sentinel, April 28, 2003, p.19A.

13. Press Release, Governor Attends Florida Fruit and Vegetable Association 60th Anniversary Meeting. Florida Department of Agriculture and Consumer Services,09–23–2003. www.doacs.state.fl.us/press/2003/09232 003.html.

14. Davis, S.,Leonard, J.B. The ones the law forgot: children working in agriculture. Farmworker Justice Fund, Washington, D.C. 2000, cited in Das,R., Steege, A., Baron, S., Beckman, J., Harrison, R. Pesticide-related illness among migrant farm workers in the United States. International Journal of Occupational Environmental Health. 2001; 7:303–312.

15. Steege, A. Unpublished 1999 data, National Agricultural Workers Survey, NIOSH Health Supplement, cited in Das,R., Steege, A., Baron, S., Beckman, J., Harrison, R. Pesticide-related illness among migrant farm workers in the United States. International Journal of Occupational Environmental Health. 2001; 7:303–312.

16. Arcury, T.A., Quandt, S.A., Cravey, A.J., Elmore, R.C., Russell, G.B. Farmworker reports of pesticide safety and sanitation in the work environment. American Journal of Industrial Medicine. 2001; 39:487–98, cited in Das,R., Steege, A., Baron, S., Beckman, J., Harrison, R. Pesticide-related illness among migrant farm workers in the United States. International Journal of Occupational Environmental Health. 2001; 7:303–312.

17. ibid

18. Mills, Paul K. and Kwong, Sandy (2001). Cancer Incidence in the United Farmworkers of America (UFW), 1987–1997. American Journal of Industrial Medicine, 40:596–603.

19. Morbidity and Mortality Weekly Report, 48 (6), 1999 cited in Mills, Paul K. and Kwong, Sandy (2001). Cancer Incidence in the United Farmworkers of America (UFW), 1987–1997. American Journal of Industrial Medicine, 40:596–603.

20. French, Kimberly. Taking Justice to the Community. UU World, September/October 2002.

21. US DOL Report to Congress: The Agricultural Labor Market—Status and Recommendations. 2000, cited in Das, Ruali; Steege, Andrea; Baron, Sherry; Beckman, John; Harrison, Robert. Pesticide-related Illness among Migrant Farm Workers in the United States. International Journal of Occupational and Environmental Health, 2001; 7:303–312.

22. Methyl Bromide, CH3Br, www.c-f-c.com/sepcgas_products/methyl-bromide.htm.

23. Methyl Bromide Questions & Answers, U. S. Environmental Protection Agency,www.epa.gov/ozone/mbr/qa.html.

24. Methyl Bromide Patient Information, Agency for Toxic Substances and Disease Registry (ATSDR) at www.atsdr.cdc.gov/MHMI/mmg16.html.

25. Methyl Bromide Questions & Answers, U. S. Environmental Protection Agency,www.epa.gov/ozone/mbr/qa.html.

26. US Department of Labor, OSHA Safety and Health Topics: Methyl Bromide, on www.osha. gov/dts/chemicalsampl ing/data/CH_251900.html.

27. Methyl Bromide, OSHA, US Department of Labor, www.osha.gov/dts/sltc/methods/partial/pv2040/2040.html, p. 2.

28. Methyl Bromide Questions & Answers, U. S. Environmental Protection Agency,www.epa.gov/ozone/mbr/qa.html.

29. Bartlett, Sarah; Hickman, John. "Critical Use" Exemptions and the Methyl Bromide Blues. Berry College. Synthesis/Regeneration 32. Fall 2003, reprinted in www.greens.org/s-r/32/32–15.html.

30. State of California Department of Pesticide Regulation, cited in Farmworkers lack aid in pesticide hits. Wire Report, Daytona Beach, Florida News Journal, May 17, 2004.

31. Secondhand Pesticides: Airborne Pesticide Drift in California, Summary of a report by Susan Kegley, Anne Katten and Marion Moses, M.D. in Global Pesticide Campaigner, Vol 13, Number 1, April 2003. Pesticide Action Network North America.

32. ibid.

33. Kirkpatrick, David D. EPA halts Florida test on Pesticides. New York Times, 4/9/05, www.truthout.org/docs_2 005/040905Z.shtml

34. Email Juliann Sum, JD, Sc.M., Coordinator of Public Programs, Labor Occupational Health Program, University of California, Berkeley, to this writer, 6/15/04.

35. Peiris-John, 2002, cited in *Chronic Neurological Effects of Pesticides.* Marion Moses, M.D., Pesticide Education Center, San Francisco, CA., 2003.

36. United States Sugar Corporation vs. G. J. Henson, Supreme Court of Florida, SC01–1127, June 6, 2002.

37. Tamen, Joan Fleischer. Ex-sugar worker wins pesticide battle. South Florida Sun-Sentinel, p.1D, June 7, 2002.

38. Needleman, Herbert L. Lead levels and Children's Psychologic Performance. New England Journal of Medicine, July 19, 1979, 163., cited in Davis, Devra. *When Smoke Ran Like Water.* Basic Books, New York, 2002, p.126–133.

39. Davis, Devra. *When Smoke Ran Like Water.* Basic Books, New York, 2002, p. 133.

40. Pesticides News, p.1,2, undated, www.pan- uk.org/pestnews/pn38p9.htm.

CHAPTER 4

▼

DAVID LEE YOUNGBLOOD: SERVED HIS COUNTRY, DIED AT THE COMPANY

David and Martha Youngblood had been married almost 20 years. "We were partners, we did everything together." The couple lived a simple life, Martha Youngblood explains, dinner and a movie was their night out. David and Martha enjoyed day trips to Branson, to explore the caves. They shared an enthusiasm for history, civil war sites and re-enactments, Victorian home tours. "All of a sudden, I have no one to do these things with. We did everything together."

Martha describes David Lee Youngblood as "a good man, a nice guy. He spoiled me, he was very attentive to me. We had no kids, we had each other."

David was born at Camp Lejeune, North Carolina, where his father was in the Marines. He was raised in the Clearwater-St. Petersburg area of Florida. After high school, he served fifteen years in the United States Air Force. His first assignment was fifteen months in Turkey, which he and Martha both enjoyed. Then they were relocated to Georgia for three years, followed by Whiteman Air Force Base in Missouri. They liked living in Missouri. David took early retirement from the Air Force and began work there in the private sector.

In October 1997, Youngblood signed on with Hayes Lemmerz International, at their wheel plant in Sedalia, Missouri. He started as a Manufacturing Attendant, and was later promoted to Manufacturing Controller. Around January 1999, he began working on the disc spinner machine, the position he held until his death.

Hayes Lemmerz International, Inc. described itself as "one of the leading suppliers of automotive and commercial highway wheels, brakes, suspension, structural and other light weight components to the global transportation industry."[1]

We know that David Youngblood was impeccably reliable: He never missed a day of work in the time he worked for Hayes Lemmerz. David would have been 43 years old in a few weeks, when the life was crushed out of him at the Sedalia plant of Hayes Lemmerz International.

David and Martha Youngblood had been married almost half their lives by that fateful evening of June 4, 1999 when David left for his 3rd shift job at the Sedalia plant as usual. But this night, he was never to come home again.

At sometime after 9 pm, David Youngblood left the control panel one flight up, came down and entered the huge disc spinner machine to make an adjustment, or remove a jam in the machine. The gate would malfunction: it should not have allowed him to enter the machine without first locking out all energy sources to the machine. It didn't do that. And David Youngblood had the life crushed out of him.

David Youngblood's upper body was pinned between two parts of the machine. His hat was found on the ground next to the transfer beam, his flashlight a few feet away. The OSHA report would state that David Youngblood was "caught pinned between the transfer beam and the orient station on a Disc Spinner Machine ..."[2]

When the ambulance and emergency personnel arrived, they found the "floor covered in oil," creating a hazardous condition, affecting their ability to provide emergency medical services. They also noted that no one had performed, or attempted to perform, CPR.[3]

"I was, and still am, heartsick when I read in the ambulance report 'negative for bystander CPR' .... Every nurse knows severe brain damage sets in after five minutes of no oxygen/blood supply to the brain. At least ten minutes went by from the time the ambulance got the call until the ambulance arrived there," Martha Youngblood would recall later.[4]

David Youngblood was pronounced dead at 11:41 pm at the Bothwell Regional Health Center. He was dead by the time the ambulance transported him there. He died of "traumatic asphyxiation," according to the death certificate.[5] Martha Youngblood sees it differently: "Hayes Lemmerz murdered my husband to make a profit. They put profit before safety."[6] The toxicology report was negative and the autopsy ruled out any health problems that might have contributed to his death.[7]

Occupational Safety and Health Administration of the U.S. Department of Labor (OSHA) was called, as required by law, and came to the plant the next day, Saturday, June 5, 1999 to begin their investigation.

Over the next several days, the OSHA investigator, would hold interviews at the Hayes Lemmerz Sedalia plant, view the scene of the accident, videotape the scene, and the machine, as well as the access gates to the disc spinner machine.

The massive disc spinner machine is surrounded by a 5–6' high fence, and has five access gates. In order to operate the machine, the manufacturing controller (David's title) operates the machine from a control panel on the mezzanine, one flight up. When the machine malfunctions, the manufacturing controller must come down, open a gate, which is supposed to lock out all energy sources, and then fix the malfunction. This is written in the job description as an "occasional duty" for the manufacturing controller, but as we shall see later in the Pass Down Logs, it appears to be a frequent occurrence.[8]

According to the OSHA tape, at around 10:55 PM, a co-worker realized that no one had seen David for some time. When he got to the walkway, the supervisor could see David slumped over the transfer beam on the machine. He opened the door and went in. He put a hand on David's back, but as he described it, David felt cool to the touch. He alerted a guard to call an ambulance immediately, then called the manager. At the same time, another worker was hitting the emergency buzzer to get more help. When they looked at the southwest (SW) gate, it was open at the time. That gate had a switch that was disabled, a supervisor added, because the door had been damaged and could not be shut.[9] Specifically, "the southwest Protective Barrier Gate interlock was found defective in that the male interlock key, that inserts into interlock, had been removed from gate door and inserted into the interlock, thus allowing employees to enter the Protective Barrier without shutting the machine down."[10]

The OSHA investigator then queries re visual inspections, "The operator does them", and weekly checks, which a supervisor reportedly does and for which he has records. "Does the operator have a checklist of items?" "I don't believe they do. I think it's just something they know to check" is the response. "Would the inner locks for the gates be checked?" "Yes, should be checked."

"Sometime back," which turns out upon probing to be January 1999, someone "told me about the interlock (referring to the malfunctioning gate). At the time, a few gates had the switches taken out of them." A supervisor then adds that he himself put all the switches on the doors, but that door he couldn't do because it was damaged. He adds that he has shown that on each of his Safety Inspection Checklists.[11]

Later, the documents are produced and it turns out that he has reported that malfunction on at least 13 occasions in the four and one half months between January 19, 1999 and David's death on June 4, 1999. Specifically, "the weekly self-inspection records show that they were having problems with the southwest protective barrier gate on 1–19–99, 2–1–99, 2–8–99, 2–22–99, 3–9–99, 3–15–99, 3–22–99, 4–6–99, 4–27–99, 5–4–99, 5–17–99, 5–26–99, and 6–1–99." No work orders could be found. The final notation three days before David Youngblood's death states, "S06 disc spinner same."[12]

In his Findings of Fact and Rulings of Law, the Administrative Law Judge summarizes the circumstances leading up to David Youngblood's death: "... sometime in early January 1999, Employer's maintenance department was doing some work on the machine, which required a portion of the machine to be lifted up by the plant's overhead crane system. In this process, the crane also hooked a portion of one of the five gates ... referred to as the 'southwest gate' ..., causing the gate to be out of alignment, and preventing the gate from shutting, and preventing the proper use of the interlock key. The gate was not repaired; instead, someone inserted the male interlock key into the interlock with the southwest gate open. This allowed the machine to continue to run without the southwest gate being repaired, but also allowed personnel to enter the protective barrier, through the southwest gate, without the machine shutting down automatically (as it was designed to do)."[13]

"... [O]n or about January 7, 1999, David Youngblood informed _____that the southwest gate had been damaged in the manner described above. Youngblood also informed_____ that the other four gates had been tampered with to allow personnel access to the interior of the protective barrier, without the machine shutting down._____ testified that he immediately asked maintenance to fix all five gates. _____testified that when the gates had not been fixed within a day or two, he fixed the other four gates himself, but he could not fix the southwest gate._____ testified that he orally requested the southwest gate to be repaired on several occasions.... ... it is clear that the southwest gate was not repaired at any time between its being damaged in early January 1999, and June 4, 1999, when David Youngblood was killed. Likewise, it is clear that the southwest gate was not chained and locked to prevent its being used pending its repair. It is also clear that, had the male interlock key not remained in the interlock of the southwest gate (with the southwest gate open) for these five months, the machine could not have run, and production would have been halted...."[14]

The OSHA investigator asks, "Was there any instruction to weld that door shut or chain it shut, so no one could get through there?" No, is the response.

Later on the OSHA tape, they talk about the latch. OSHA investigator surmises that if the latch is not holding in the proper position, and wasn't making contact, then the machine wouldn't run. Therefore it had to be jerry-rigged, or bypassed, in order to make the machine work. This assumption is acknowledged in the affirmative by the Hayes Lemmerz staff member.[15]

The Sedalia plant's policies state: "Department Managers and Supervisors will monitor the safety, health and environmental programs in their respective areas.... machines or equipment shall be operated only with the proper or designed safety devices in place. All equipment is to be operated in a safe manner and all safe operating procedures such as lockout/tagout ... shall be followed."[16]

During the ensuing days on site, several safety violations are identified by OSHA:

"The floor inside, and outside the protective barrier around the machine was coated with oil, making it hard to walk without slipping and falling, creating a hazard to the Manufacturing Controller of the machine and Maintenance men that work in this area." It is further noted that the employer "with reasonable diligence ... should have known.... [W]hen we were in the area they advised me to be careful, the floor was slippery."[17]

Employees were "exposed to hazardous conditions, due to the company Energy Control Program, Lockout/Tagout found to be inadequate because the machine having more than one energy source.... (and) ... program "did not have specific procedural steps for each different type of machine or equipment for shutting down, isolating, and securing machines or equipment to control hazardous energy sources.... (nor) specific requirements for testing a machine or equipment to determine and verify the effectiveness of the lockout devices." The OSHA inspector then observes a machine upon which maintenance is in process "and the lockout/tagout procedure they were using was a tag on the off and on switch, on the machine only." The employer "should have known, the plant lockout/tagout program was approved by six different area managers ..."[18]

Lockout refers to the "placement of a lockout device on an energy isolating device ... ensuring that the energy isolating device and the equipment being controlled cannot be operated until the lockout device is removed."

Tagout refers to the "placement of a tagout device on an energy isolating device ... to indicate that the energy isolating device and the equipment being controlled may not be operated until the tagout device is removed."[19]

The Hayes Lemmerz Sedalia plant procedure for Lockout-Tagout, states its purpose is "to assure safe shut down of equipment by blocking, stopping, and reducing equipment to the zero mechanical state when performing work on electrically controlled equipment." The Company's procedure also states that all employees be issued a padlock.[20]

The reality appears to be different: "Locks, tags, chains, wedges, key blocks, adapter pins, self-locking fasteners, or other hardware were not provided by the employer for isolating, securing, or blocking of machines or equipment from energy sources: The employer did not provide the manufacturing controller ..., the hardware to lockout and tagout all energy sources to the machine before entering the protective barrier area for adjustments, replacement of parts, and removal of falling or jammed parts." Yet the manufacturing controller and others have to "enter the protective barrier area to make adjustment or replace part, and remove jams, and that is part of the operator duties." "The likelihood an injury will occur is judged to be relatively high, because they enter the protective barrier by pushing the emergency stop and not de-energizing all the energy sources."[21]

An additional violation is the lack of adequate training in the company's "lockout/tagout program for all the energy sources on the machine and the locations of the energy sources to be lockout to render the sources isolated." It also notes that, "by using the emergency stop to shut the machine off, the emergency stop only shut the machine out of the automatic mode."[22]

The disc spinner machine was shut down at lunch-time (8:00 P.M.) on June 4, 1999. The manufacturing controller (David) was told there was no need to run the machine, "unless he wanted to feed parts manually." The supervisor states it is possible to run the machine by oneself.[23]

Pass Down Logs are used to document problems and corrective actions that occurred during each shift. The Pass Down Logs for the manufacturing controllers/operators demonstrate that it is part of their normal duties to enter the protective barrier area to make adjustments, remove jams, fix, replace, etc.[24]

Pass Down Logs are also a tool to record and monitor each manufacturing controller's hourly progress toward his or her production quota. "These machine operators have a quota they are to meet on each machine, and on occasion the manufacturing controllers will operate the machine by them (sic) self."[25] Both the daily shift Pass Down Logs and internal memos appear to bear out this pressure on production. Minutes from a department meeting note "S06 still needs to run 3000 per shift...."[26]

The manufacturing controller's job description notes that the controller may "occasionally" need to enter the protective barrier gate to remove jams, make adjustments, etc. The reality, as documented in the Pass Down Logs, appears quite different.

We reviewed the Pass Down Logs for all three shifts for the 42-day period up to, and including, the night of David Youngblood's death. Fully two-thirds of shifts appear to have identified one or more problems on the machine. The litany included broken bolts, bolts stuck in the machine, stuck parts, loose parts, broken rollers, broken hoses, hydraulic leaks, broken pumps, burned out motors. These problems, on average, appeared to be a daily occurrence and engaged the manufacturing controller's time at the same time that they had to produce to meet the quota. [27] The down time caused by stopping to fix these problems was likely to create increased pressure.

We know from David Youngblood's Pass Down Log that he was alive after 9 PM: that's how far his Log was completed for June 4th. We will never know exactly what happened, but one can conjecture that the machine stopped because something had jammed it. David came downstairs and opened the Southwest gate, thinking that the machine was automatically in lockout mode once he had opened the gate. It was not. Thus, when he removed the jam or made the adjustment that was stopping the machine, it started up again and crushed him. From the position of his hat, with his flashlight and his wrench nearby, we can assume that he believed he was working on a stopped machine. His upper body was crushed between the heavy transfer beam and the orient station when the machine re-started.

When OSHA completed its inspection and investigation, it concluded by citing Hayes Lemmerz International, Inc. at its Sedalia, Missouri plant for four Serious and one Willful violation as follows:[28]

Citation 1 Item 1: Serious

(a) ... "The floor inside, and outside the protective barrier around the machine was coated with oil, making it hard to walk without slipping and falling, creating a hazard to the Manufacturing Controller of the machine and Maintenance men that work in this area."
Proposed Penalty
~~$3500.~~ $2800.

Citation 1 Item 2: Serious

(a) "The employer Energy Control Program (lockout/tagout) procedures were found to be inadequate...."
Proposed Penalty
~~$3500~~.$2800.

Citation 1 Item 3: Serious

"Locks, tags, chains, wedges, key blocks, adapter pins, self-locking fasteners, or other hardware were not provided by the employer for isolating, securing, or locking of machines or equipment from energy sources:
(a)The employer did not provide the Manufacturing Controller, ... the hardware to lockout and tagout all energy sources to the machine before entering the protective barrier area for adjustments, replacement of parts, and removal of falling or jammed parts."
Proposed Penalty
~~$2500~~.$2000.

Citation 1 Item 4: Serious

"The employer did not provide adequate training...."
Proposed Penalty
~~$3500~~ $2800.

Citation 2 Item 1: Willful Reclassified to Section 17

"Machine guarding was not provided to protect operator(s) and other employees from hazard(s) created by point-of-operation, ... and rotating parts, (a) ... The Protective Barrier around The Hess ... (Disc Spinner Machine) ... The Southwest Protective Barrier gate interlock was found defective in that the male interlock key, that inserts into interlock, had been removed from gate door and inserted into the interlock, **thus allowing employees to enter the Protective Barrier without shutting the machine down."** (bolding added).
Proposed Penalty
~~$56,000~~.$44,800.

Willful is the most serious classification that OSHA can give. "A violation is considered willful by OSHA if committed with intentional, knowing or voluntary disregard for the requirements of the Occupational Safety and Health Act of

1970 (the OSHA act ...), or with plain indifference to employee safety and health."[29]

OSHA Informal Conference Notes, assumed date 11/23/99 state, "I explained that the Willful was based on the fact that the interlock was reported to Maint. and Supervisors numerous times and was never completely corrected. After a discussion with the Deputy RA (Regional Administrator) we offered them to reclassify the Willful to a Section 17 and reduce the penalties by 20% with total penalty of $55,200. The ISA (Informal Settlement Agreement) was signed." Of that final total penalty amount, $44,800 was for the Willful, reclassified to Unclassified citation above.[30]

In a letter written by Charles Adkins, OSHA Regional Administrator, in response to a letter from Senator Paul Wellstone; Mr. Adkins justifies why the Willful citation was reclassified:

"the necessary tests of a high level of knowledge on the part of the employer of the standard violated and a deliberate state of mind to violate the standard would not likely be sustained through the legal process. Therefore, for settlement purposes, the citation was sustained but re-characterized without classification.... Had OSHA proceeded with litigation against the company there was a possibility, although small, that the court would rule against the Agency in a subsequent hearing. In that event, the court could not only reclassify the violation but vacate the citation altogether thereby leaving current employees exposed to the hazard."[31]

"It is a federal crime for an employer to cause a worker's death by willfully violating safety laws." However, federal law considers it a mere misdemeanor with a maximum sentence of six months.[32]

The operative word in the above paragraph is "willfully." What happened in David Youngblood's case is typical. Under pressure from the employer, OSHA reduces a "willful" violation to unclassified, thus precluding any referral to United States Department of Justice for prosecution. The assumption is that OSHA can move the company swiftly to the safety improvements in the OSHA citation. As Martha Youngblood can attest, families are not consulted by OSHA on such decisions.

The sad reality is that from 1990–2002, only 3.9% of federal OSHA willful violations resulting in deaths have been prosecuted in federal courts. In states with state-run OSHAs, only 4.6% of willful violations resulting in worker deaths have been prosecuted. Only California seems to have aggressively sought criminal prosecution in willful violation death cases.[33]

As Martha Youngblood sees it, "Hayes Lemmerz *willfully* did not shut that machine down. And *willfully* did not repair that safety gate. And *willfully* told my husband to use that machine in spite of that gate. It is disturbing to know that it took the death of an employee to finally get that machine shut down and the gate fixed.... Greed killed my husband."[34]

David Lee Youngblood died an agonizingly painful and untimely death. For Hayes Lemmerz, an international corporation with 40 plants, 11 joint venture facilities, and sales of $2.5 billion,[35] the final monetary penalty for months of failure to repair the gate, failure to provide proper lockout/tagout devices, failure to train on proper procedures was $55,200.

Hayes Lemmerz International attributed culpability thusly: "not as a disregard for Safety by the Sedalia operations, but rather a failure of the management system in place at the time." [36] One must ask how many levels of management have to be involved before a corporation takes responsibility for its actions?

And the failure that was identified so repetitively over so many months?

An internal Hayes Lemmerz memo dated June 6, 1999, two days after David Youngblood's death, states: "The north spinner on the southwest door needed more work. Door was out of square.... The east leg were (sic) the latch is at (sic) was bolted to a piece of grating. The grating moves about one inch and the door could never have latched correctly. We pulled the leg over to the correct position and welded grating down plus leg to stairway. The plunger was reinstalled and align (sic).... the safety switches need check out by Ohm meter only. The power to this machine at no time is to be turn (sic) on till OSHA is here to see machine in operation."[37]

In two days, the gate was fixed. "Why didn't they do this before my husband's death?" asks Martha Youngblood.

In a letter from Hayes Lemmerz to OSHA, Kansas City, Missouri, dated May 30, 2000, it is stated that the corrective action steps that the corporation has taken to abate the OSHA citations have cost "a $1.8 million capital investment" and the letter adds that Hayes Lemmerz now has "a world class lockout/tagout program for our employees to use."[38]

A photo diagram included in the company's submitted documents states that the "old disconnects were on the mezzanine on the next level and the procedure required six locks and a bleed down of the hydraulic and pneumatic systems. The new lockout/tagout disconnects are located at the operator station and require only two locks for the procedure ..." The new lockout center includes a group lockout box, a laminated copy of the machine specific procedure, a schematic of the lockout points and gang locks."

In February, 2001, the Circuit Court of Pettis County, Missouri, issued its ruling in the Defendant's (Hayes Lemmerz) Motion to Dismiss the Wrongful

Death Action brought by David Youngblood's widow, Martha. The Wrongful Death Action was brought on the basis that the death was "'willful and intentional' and was committed with the knowledge that such conduct of the defendant was 'substantially certain' to result in injury and/or death ..., and the fatal injury resulting to Mr. Youngblood was, therefore, not 'accidental' under the Missouri Workers' Compensation Act...." The Court determined that whether the injury and death are the result of accident or intentional conduct is the exclusive jurisdiction of the Missouri Labor and Industrial Relations Commission, and that the Circuit Court has no jurisdiction.[39]

In the previously referenced evidentiary hearing in the case held before an administrative law judge in the Department of Labor and Industrial Relations of Missouri on January 14, 2002, the issue of whether David Youngblood's death is an accident or an intentional act is explored:

"Employer failed to act by (1) Failing to repair the southwest gate immediately, and/or (2) Failing to chain and lock the southwest gate until it could be repaired, and/or (3) Failing to shut down the machine until the southwest gate could be repaired.... Employer knew the southwest gate was damaged and open for five months, and knew the machine was running nonetheless..... ... Employer's failure to act was 'intentional.'"

But "Was Employer 'substantially certain that injury to an employee will result' from its failure to act?" the judge ponders in making his decision.... "Employer's conduct in allowing this dangerous condition to exist for five months is inexcusable. It may justly be characterized as reckless or wanton," ... but finds "Employer could not have been 'substantially certain' that Youngblood would be thus injured."[40]

Thus it is the judge's decision that the exclusive remedy is the State of Missouri Workers' Compensation Law, specifically medical expenses, burial expense not to exceed $5000, and weekly compensation equal to two-thirds of her deceased husband's salary, up to 105% of the state's average weekly wage, less than $600/week.[41] A 15% penalty was added.[42]

In State of Missouri, when the injury "is caused by the failure of the employer to comply with any statute in this state ..., the compensation and death benefit shall be increased by fifteen percent."[43] In this case, the employer did not comply with 292.020, Revised Statutes of Missouri, which states in part, "machines, machinery, gearing and drums in all manufacturing ... establishments in this state, when so placed as to be dangerous to persons employed therein ... while engaged in their ordinary duties, shall be safely and securely guarded when possible; if not possible, then notice of its danger shall be conspicuously posted ..."[44]

In a letter to John P. Madigan, Jr., Acting Chairman, Labor and Industrial Relations Commission, filed April 10, 2002, Martha Youngblood writes, "In defense of my husband, David Lee Youngblood, who can no longer speak for himself and defend himself ...:

"My husband suffocated to death because a machine he should have never been allowed to use crushed his chest.

"I will always believe my husband was trained to use that machine with that gate broken and the safety interlock tampered with and altered to be ineffective."

... "If Dave was not supposed to use that gate, why didn't Hayes Lemmerz chain or weld that gate shut?.... To this day Hayes Lemmerz has not answered that question.

... "(an individual in a supervisory capacity) told the OSHA investigator (on video) the reason the interlock was tampered with was 'to keep production going.' I think Hayes Lemmerz put profit before safety.

"OSHA..asked one of the managers why it took so long to get that gate fixed. To this day Hayes Lemmerz has not answered that question. I would like an answer.

... "I want to know why Hayes Lemmerz did not care about work place safety.

"I want to know why Hayes Lemmerz did not comply with OSHA safety regulations BEFORE my husband's death.

"I want to know why Hayes Lemmerz decided to cooperate with OSHA AFTER my husband's death. Why did they wait for Dave's death to occur before they would fix that gate ...?"[45]

At the end of the Hayes Lemmerz International's Vision Statement in 2004 is the phrase: "Never Compromise: Integrity, Safety, Quality, or Customers!"[46]

References:

1. Award-winning Commitment to Quality Manufacturing at Hayes Lemmerz is recognized by Industry Leaders. Hayes Lemmerz News Release, dated May 26, 1999, www.hayes-lemmerz.com/press..tml/1999/19990526_qualitymfg.html.

2. OSHA Inspection No. 302247218, Investigation Summary, dated July 28, 1999.

3. Missouri Ambulance Reporting System Run Report No. 9917559, Ambulance Service No. 159042.

4. From Martha Youngblood's handwritten note on invoice from the medical response group that responded on 6/4/99. Note: This bill was paid by Martha Youngblood after Hayes Lemmerz' insurer did not appear to have paid it within three months. It was later reimbursed to her by the insurer.

5. Missouri Department of Health, Certificate of Death for David Lee Youngblood, signed 6/10/99, filed 6/18/99.

6. Letter from Martha Youngblood to United States Senator Paul Wellstone, September 11, 2002.

7. Letters from Pettis County Coroner to Martha Youngblood, 7/23/99 and 8/18/99.

8. Hayes Lemmerz International Employee Classification Profile, Manufacturing Controller, in OSHA Inspection No. 302247218.

9. OSHA Inspection No. 302247218, Interview tapes, June 5–9, 1999.

10. OSHA Inspection Number 302247218,Citation 2, Item 1, revised citations, Informal Conference, November 23, 1999.

11. OSHA Inspection No. 302247218, Interview tapes, June 5–9, 1999.

12. Hayes-Lemmerz International Self Inspection Checklists, in OSHA Inspection No. 302247218, and Citation No.02, 001, p.16.

13. Findings of Fact and Rulings of Law: Injury No. 99–068464, before the Division of Workers'Compensation, Department of Labor and Industrial Relations of Missouri, dated January 23, 2002, p.5.

14. ibid.

15. OSHA Inspection No. 302247218, Interview tapes, June 5–9, 1999.

16. Hayes Lemmerz Sedalia Plant Policy on Safety, Health and Environment, POL00035–03, Effective November 23, 1998, in OSHA Inspection No. 302247218.

17. OSHA Inspection Number 302247218, Worksheet, Citation 1, Item 1.

18. OSHA Inspection Number 302247218, Worksheet, Citation 1, Item 2.

19. OSHA Regulations, 29 CFR 1910.147, The control of hazardous energy (lockout/tagout).

20. Hayes Wheels Sedalia Plant Procedure Lock Out—Tag Out, Effective December 11, 1997, in OSHA Inspection No. 302247218.

21. OSHA Inspection Number 302247218, Worksheet, Citation 1, Item 3.

22. OSHA Inspection Number 302247218, Worksheet, Citation 1, Item 4.

23. OSHA Inspection Number 302247218, Hayes Lemmerz International, Inc. Accident Summary.

24. OSHA Inspection Number 302247218, Worksheet, Citation 1, Item 4.

25. OSHA Inspection Number 302247218,Accident Summary.

26. Minutes, Hayes Lemmerz Department Meeting, S06-S07, dated 1/7/99, in OSHA Inspection No. 302247218.

27. Hayes-Lemmerz Pass Down Logs, April 24, 1999 through June 4, 1999, in OSHA Inspection No. 302247218.

28. OSHA's Citation and Notification of Penalty for Hayes Lemmerz International, Inc., Sedalia Missouri, Inspection No. 302247218, Inspection dates 6/05/99–10/28/99, Issuance Date: 11/04/99.

29. Letter from Charles Adkins, Regional Administrator, OSHA Kansas City, Missouri to United States Senator Paul Wellstone, August 27, 2002.

30. OSHA Inspection Number 302247218, OSHA Informal Conference Notes, 11/23/99.

31. Letter from Charles Adkins, Regional Administrator, OSHA Kansas City, Missouri to United States Senator Paul Wellstone, August 27, 2002.

32. Barstow, David. When Workers Die: The Plumber's Apprentice. New York Times, December 21, 2003, p. A1.

33. Barstow, David. When Workers Die: The California Way. New York Times, December 23, 2003, p. Al.

34. Letter from Martha Youngblood to Ron Cucuz, CEO, Hayes Lemmerz dated July 5, 2000.

35. Hayes Lemmerz News Release, dated May 26, 1999, www.hayes-lemmerz.com

36. Letter from Hayes Lemmerz International to Lodama Delinger, Area Director, OSHA, Kansas City, Missouri, undated, in OSHA Inspection Number 302247218.

37. Hayes-Lemmerz Internal Memo, dated June 6, 1999, in OSHA Inspection Number 302247218.

38. Letter from Hayes Lemmerz to OSHA, Kansas City, Missouri, dated May 30, 2000, in OSHA Inspection No. 302247218.

39. Youngblood v. Hayes Lemmerz International, CV400–754CC, Ruling on Defendants' Motion to Dismiss, dated Feb. 14, 2001.

40. Findings of Fact and Rulings of Law: Injury No. 99–068464, before the Division of Workers' Compensation, Department of Labor and Industrial Relations of Missouri, dated January 23, 2002, pp.6,7.

41. ibid and Revised Statutes of Missouri, Chapter 287, Section 287.240.

42. Findings of Fact and Rulings of Law: Injury No. 99–068464, before the Division of Workers' Compensation, Department of Labor and Industrial Relations of Missouri, dated January 23, 2002, p.8.

43. ibid, and Revised Statutes of Missouri, Chapter 287, Section 287.120.4.

44. Revised Statutes of Missouri, Chapter 292, Section 292.020.

45. Letter from Martha Youngblood to John P. Madigan, Jr., Acting Chairperson, Labor and Industrial Relations Commission, Missouri Department of Labor and Industrial Relations, date filed April 10, 2002.

46. http://www.hayes-lemmerz.com/about/html/our_vision.html, ©1998–2004

CHAPTER 5

▼

HAYES LEMMERZ 2

On October 29, 2003, at another Hayes Lemmerz plant, this one in Huntington, Indiana; Shawn Duane Boone met an excruciating, agonizing death as his body, skin, hair, and internal organs burned in an aluminum dust explosion. This time, there was not even any specific OSHA regulation governing the cause of his death, aluminum oxide explosion. Nor was this the first death at the Huntington plant.[1]

Shawn Boone loved the rural life, enjoyed fishing, and his dog, Dutchess. He had just bought a farmhouse. Tammy Miser speaks glowingly, lovingly of her baby brother, Shawn: He was a "can do" guy with a sense of humor, well liked by all—even former girlfriends. Miser says Shawn loved to tinker and to fix things. He had an enormous tractor, not very good, but he would get it out to give rides to his nieces and nephew when they came to visit. The kids loved it. "Shawn was real lovable," Tammy remembers, a good brother, son, friend, and beloved uncle to Tammy's three children. "He made the children feel special." Now his remains lie buried in a nearby cemetery. Shawn Boone was 33 years old.

Boone was taking courses toward a degree at a technical college. He had been a maintenance technician and mechanic at the Hayes Lemmerz plant in Huntington, Indiana for four years when he died. The plant makes aluminum alloy wheel rims for cars.

Aluminum dust is highly explosive. Tammy Miser says her brother was right in front of the furnace, picking up his tools, when it happened.

The furnace is used to melt the aluminum chips that are left over after machining the aluminum wheels cast at the plant.[2]

Earlier in the day, about 3:00 pm, there had been a fire in the same number 5 furnace chip melt well exhaust duct. The maintenance department evaluated the fire and shut down the exhaust ventilation system to allow the fire to self-extinguish, "as this was the plant policy for aluminum dust fires in the pipe. The employees were then to clean out the pipe and restart the system."[3]

About 8:15 pm after the pipe had cooled and been cleaned out, Boone and two other workers were given instructions to start the system back up again. Ten to fifteen minutes later, the explosion occurred.[4] A mechanic noticed "chips falling from the spark box. He turned to tell his coworkers to stop the dry chip feed. As he turned, a fireball erupted from beneath the furnace fume hood. The fireball ignited his clothing and totally engulfed another mechanic (Shawn Boone) standing near the sidewell vortex pump. The flames also singed a third mechanic, who was standing near the system control panel."[5]

The explosion rattled windows more than a mile away. The Huntington Fire Chief was reported to have seen the fireball from his home about eight miles away.[6] The roof was blown off and some of the side wall as well. The door to number 4 furnace was blown 50 feet.[7] A pair of safety glasses found afterwards near the number 5 furnace had molten metal on the front. A burned watch and burned hat and a pocketknife with burned material, possibly clothing were also found nearby.[8]

The Huntington Fire Department responded to the scene. Two firefighters, with the help of employees, carried Shawn Boone out of the building on a backboard.[9] Miser describes how her brother did not die right away, but he was still conscious. He was "smoldering while the aluminum dust continued to burn through his flesh and muscle tissue. The breaths that he took, while in the building, burned his internal organs." The only skin left was some on the side of his face that was facing away from the furnace, she explains. He asked for help and something for the pain. He was airlifted to a hospital in Ft. Wayne. Miser drove five hours to see him. The hospital told her Shawn couldn't survive. His internal organs were all burned. The heat from the blast was so hot it melted the copper piping on the roof, she adds.

No one had given his name to the fire or emergency personnel. Thus when Shawn arrived at the hospital, Mark Miser, Tammy's husband and Shawn's brother-in-law had to identify him by telephone. Tammy reports that Mark started by describing his red hair, but Shawn's hair wasn't there anymore.

Tammy couldn't believe that her younger brother couldn't be saved. When the hospital told her there was nothing else they could do for Shawn, Tammy Miser had to make the terrible decision that no family member ever wants to have to make for a loved one: to take him off life support. Shawn Duane Boone's life gave out early on the morning of October 30, 2003. He had endured third degree burns over 95% of his body.[10]

How did this happen?

According to the Material Safety Data Sheet produced by the manufacturer, "suspensions of aluminum dust in air may pose a severe explosion hazard. A potential for explosion exists ... if at least 15% to 20% of the material is finer than 44 microns (a micron equals one millionth of a meter) .... Molten aluminum may explode on contact with water."[11]

"Fine particles of aluminum powder ... are easily dispersed in air where their low mass allows them to remain suspended or 'float' in the air." The particles will burn when their ignition temperature is reached.... "[W]hen dispersed in the proper proportion in the air, which allows the particles to mix with oxygen, the burning extends from one particle to another with such rapidity, that a violent explosion results.... [V]ery little aluminum powder is needed for an explosion to occur. Aluminum powder in the form of a dust cloud will ignite with as little as 6% oxygen present....

"The discharge of static electricity will produce an electric spark that could raise the temperature of suspended powder particles in the vicinity of the spark above the ignition temperature—resulting in a fire or explosion.... [A]nything producing a spark can set off an explosion of aluminum powder suspended in air."[12]

Within the Hayes Lemmerz Huntington plant is a foundry with aluminum melting furnaces. The number 5 furnace, the area of the explosion and the one at which Shawn Boone was working, reportedly had an 80,000 pound capacity.[13]

The furnaces are attached to an aluminum chip recovery system that originates in the machining area where "aluminum chips are produced as by-products of the aluminum rims when they are machined. The chips wet with machine oils are then transported by a pneumatic system to the wet chip hopper in the number 5 furnace area. The chips then go to a rotary dryer adjacent to ... the wet chip hopper. Once the chips are dried they are then augured out of the dryer into another pneumatic system that transports the dried chips to the dry chip hopper.... From the dry chip hopper the dry chips are augured into another pneumatic system that takes them to the number 5 chip melt well at the number five furnace...."[14]

Hayes Lemmerz had an extensive history of fires in that part of the plant over several years, including a "history of aluminum dust explosions in the number 5 chip well hood which resulted in fires in the … ventilation duct work, ceiling and equipment areas.… [T]he number 4 furnace chip well system … had the same tendencies to have aluminum dust explosions upon startup."[15]

The same explosion that killed Shawn Boone, severely burned another employee and sent 5 others to the hospital for emergency care. Forty-five employees were exposed to burns, smoke inhalation and death that day.[16]

One employee describes how chips were building up on the furnace top and the thermal coupler the day before and the day of the explosion. When the system was running, hot chips were falling on him when he was in the area. Indeed IOSHA noted that "the floor areas adjacent to the number 5 chip well, rotary dryer, wet chip hopper and dry chip hopper have accumulations and piles of aluminum chips and dust around and on them. There are accumulations of aluminum chips and dust on equipment ledging as well as the electrical installations, such as but not limited to, the disconnect boxes, control panels and junction boxes. These conditions are verified by employee interviews."[17]

"It was like an earthquake", one worker recalls. Afterwards he heard three to four explosions, and could feel the building shaking.[18]

As one employee explained it, "a big flash came out of the well … this moved the dust off of the ceiling and ignited it causing a ball of fire to move through the room … started many small fires inside the building.…"[19]

The Fire Chief is reported to have said the roll out used to put out the fire was empty when they arrived. He was reportedly told they had used it on a fire earlier in the day.[20]

In one instance about six months before the fatal fire, molten metal had splashed unto the ceiling rafters, where it was smoldering. "Stuff was falling from the roof," describes a cleaning contractor who was present that day. He told a foreman that there was a major fire and he needed to call the fire department. The reply was that they were going to let it burn out.[21] Another worker described the same fire thusly: "it got so hot that the roof beams on the furnace were glowing red."[22]

At the opening of Indiana Department of Labor, Occupational Safety and Health Administration's(IOSHA) investigation, company officials are reported to have raised issues regarding the interviewing of employees. There followed a telephone discussion with Indiana's Commissioner of Labor. IOSHA stood its ground.[23]

A former supervisor at Hayes Lemmerz states there were several fires in the vicinity of the explosion, related to the "ignition of dust piles by electricity or liquid metal." Further, he states that plant management had been made aware of the situation.[24]

A former worker at the plant remembers that in March 2002, following a truck fire, the policy began that a supervisor had to be notified, who would then notify the maintenance supervisor who would inspect the fire. "At no point was (sic) we to notify the Fire Department." He recalls seeing six to eight inches of dust on furnace number five that caught on fire. He states that he reported it to managers. Their solution "was to use lime to (put it) out. I got mad, threatened to call OSHA, ... when we left the plant at the end of our shift it was still burning. One week later I was transferred out (into another unit or department), one month after that I was fired.." after he attended a family funeral.... "_____was notified at all time(s) and would never respond."[25]

The plant's fire procedures document that it is the role of the Emergency Coordinator or designee to call the Fire Department or other emergency response agencies.[26]

One employee recalls the day before the explosion thusly: "On October 28th, the burners on furnace #5 shut off. I had trouble relighting #6 burner. I found an accumulation of chips 15–18 inches deep around the burner, which had to be dug out to get to the components of the burner. While I was servicing this burner, I was being 'rained' on by chips coming out of a hole located in the side of the cyclone.... whenever I would do any servicing work in the chip melt department a dusting' of dust would accumulate on my tools and tool cart.... the entire system when in normal operation was pressurized which was forcing dust out of any crack or crevice in the system."[27]

Another employee notes that (aluminum) dust particles were in the air constantly. "If you walked back in the chip area, a fine dust would get on clothes and tools, and you could visually see the dust in the air."[28]

Other employees noted five to seven inches of dust on heat shields and several inches of aluminum dust on all electrical boxes and control panels. "The general area had only been cleaned once in a year. In the cyclone for the chip feed system, there were holes ranging from a pin to a penny. There were about a dozen or more holes in the cyclone for the chipfeed...."[29]

The cyclone above the number 5 furnace chip well had a hole described as 1 inch x 3/8 inch to ½ inch that "someone had attempted to seal with a rubber type compound that appears to be silicon caulking." This hole was positioned in such a way that it "caused the buildup of aluminum chips and dust on top of … number 5 furnace."[30]

Upon inspection, the cyclone for the dry chip hopper was found to have a hole in the top seam 4 inches long by 1/8 inch wide, and another in the bottom seam of 3 $^{1/2}$ inches by 1/8 inch that someone had attempted to seal …, as well as numerous welded patches and patches over patches."[31] Welding was thus being performed in the presence of explosive aluminum dust.

On October 24th, five days before the explosion, in repairing a leak on the cyclone of the dry chip hopper, it was necessary for the worker to remove an accumulation of dust and chips on top of the hopper. "The chips were in a pile 24 to 30 inches tall. The top of the bin was covered with dust ¼ to ½ inch deep." On October 23rd, the outside contractor had completed the cleaning of the tops of the wet and dry chip hopper, chip melt reverb, the top of number 5 furnace. As the IOSHA inspector notes, it is an indicator of the amount of aluminum chips and dust the system was putting out.[32]

One employee notes, "The cleaning out of the system … would almost always cause a small fire in the exhaust hood. On furnaces 4 and 5 … the fire would most of the time overwhelm the exhaust hood. Sometimes the fires would reach the roof of the plant. But the fires would burn out almost immediately. If the system was shut down … for a long period of time the restarting of the system would cause large flashes. While operating the chip melt system the chips were falling from a hole somewhere above the furnace…. Chips were falling in about a 4' area around the vortex pump. Just moments before the explosion I smelled an odor, which usually indicates that the chips that are being transferred into the furnace were backed (up)._____was told about the leaks…. I told them it wasn't very safe at all and I wanted them fixed."[33]

Another employee states, "There was an odor when it plugged up and it would overflow."[34]

Another employee, recalling the explosion, is cited: "He remembers the chip feed to the chip melt well was overflowing out the overflow into the chip melt area. He … remembers a sound like a fire cracker or a light bulb popping, then another louder noise, then covering up trying to protect himself from the fire."[35]

It is posited that the explosion on the night of October 29th occurred because "a small dust explosion inside the number five chip well then ignited the aluminum dust that was suspended in the air around the chip well hood due to the hole in the cyclone above the chip well. The ignition of this dust cloud then shook loose the dust that was accumulated in the truss and roof structures above that area creating much larger explosions at the roof area…."[36]

"It appears that the earlier duct fire was not the ignition source for the explosion but that another dust explosion in the chip well hood was.... .both were initiated by the chip well hood aluminum dust explosions that were common to the system."[37]

This conclusion differs from that of the Fire Marshal as stated in his report: "... It is the opinion of the investigators on the scene that the fire in the chip dryer duct work was never fully extinguished and properly cleaned, or a fire re-ignited in the duct work, perhaps from the burning catch drum. This duct fire then ignited the dust explosion that occurred in the reverb area."[38]

Analysis of samples taken from the number 5 furnace area and at the chip feed confirmed the presence of explosive material.[39]

"... it does not appear that employees were ever trained about the hazards associated with aluminum dust."[40]

The National Fire Protection Association (NFPA) has established voluntary guidelines for controlling combustible dusts and preventing explosions. The NFPA codes are the standard for combustible metals and metal dusts. NFPA 484, 2002 Edition, Section 4.4.4 Compressed Air-Cleaning Requirements states: "Compressed air blowdown shall not be permitted, except in certain areas that are otherwise impossible to clean and, where permitted, shall be performed under carefully controlled conditions with all potential ignition sources prohibited in or near the area and with all equipment shut down."[41]

It appears Hayes Lemmerz allowed compressed air to be used to clean areas around the number 5 furnace area that contained explosive aluminum dust. Maintenance used air hoses to clean things off. They reportedly were not told not to use air hoses for cleaning until after the explosion. The contract cleaner also used compressed air to clean the top of the furnace. Employees in the area would use a broom and sometimes an air wand to clean the number 5 furnace area that always had more dust and chips.[42]

NFPA recommends permanently installed vacuum cleaning systems for maximum safety "because the dust-collecting device and the exhaust blower can be located in a safe location outside the dust producing area."[43]

"The dust collector was not separate from the foundry building and personnel ... as recommended in NFPA 484. This lack of separation increased the risk of injury to personnel in the event the dust collector exploded."[44]

"There appears to be a significant housekeeping problem related to the quantities of aluminum dust and chips that were observed in the area on the floors and equipment, apparently from leaks in the system. There was a cleaning contractor hired to clean the tops of the number five furnace, wet chip hopper, dry chip hopper, and the interior of the pipe from the number five furnace chip well to the multi-clone system. No evidence could be found that would indicate the ceiling areas, ceiling trusses, tops of duct work, and piping, electrical cable shielding, and wall spaces were *ever* (italics added) cleaned."[45] "On the contrary, evidence supports the fact that the only time the roof trusses, insulation and other material were cleaned was by a previous fire."[46] The cleaning contractor tried to make the cleaning a monthly job, but says Hayes Lemmerz never signed off on that. According to the contractor, they cleaned whenever Hayes Lemmerz called them in.[47] It appears from a review of work sign-off sheets for the contractor that the frequency of cleaning was about once every two months.[48]

It is disturbing that Hayes Lemmerz had been cited previously at this plant: "The employer shall control accumulations of flammable and combustible waste materials and residues so that they do not contribute to a fire emergency."[49]

The contract cleaning company described how they would "push the pipe cleaner through pipe and step back, getting out of the way from the fire or small explosion that was about to happen."[50]

Hayes Lemmerz also dictated the placement of explosive aluminum dust and chips "back into the main process stream for processing when they placed floor sweepings that contained the fugitive aluminum dust and chips from the number 5 furnace area into the Jeffers feeder for reprocessing."[51] Thus more explosive dust was being added back into the system.

The employees that dealt with the smoldering fire that afternoon and evening were not provided personal protective equipment or fire retardant clothing, although it was known that they were dealing with an aluminum dust fire and cleanup.[52]

When working with combustible metals, NFPA codes recommend that clothing worn in the area should be "flame retardant, and non-static generating where combustible aluminum dust is present and shall be designed to be easily removed." "Safety shoes shall be static-dissipating, … shall have no exposed metal…."[53]

Hayes Lemmerz own hazard assessment performed for the Melt Operators in May 2002, and cited in the IOSHA Inspection report, lists "falling objects … hot sparks, molten metal splash, high temperatures and dust," and notes protection needed: "hard hat with Al shield," … "hot mill gloves," … "steel toed shoes," … "flame retardant jackets," … "safety glasses and face shields."[54]

Hayes Lemmerz was clearly aware that protective clothing should have been provided to workers. In fact, Hayes Lemmerz had already been cited for that violation in 1997: "The employer shall provide at no cost to the employee and assure the use of protective clothing which complies with the requirements...."[55]

"There were several fire blankets ... in the first aid room...." But it appears that not until after the explosion were "WaterJel fire blankets and pads in the furnace area ..." installed.[56]

It appears Hayes Lemmerz did not evacuate the employees in the area when there was an aluminum dust fire, contrary to NFPA standards.[57]

One of the two contractors on site to perform air emissions testing described flaming debris coming through the ceiling of their trailer and still flaming as it hit the floor. He pushed the hot door open, suffering 2nd degree burns on one hand. Outside, he saw that the baghouse, part of the dust collection system, was "completely engulfed in flames." He described it as appearing to be receiving fuel. The roof of the building was in flames and the metal mangled.[58]

NFPA standards dictate that only trained personnel shall engage in fire control activities.[59] This, too, appears to have been routinely ignored by Hayes Lemmerz. "Duct work fires were assigned to the maintenance department. Three maintenance personnel who had welded on the number five furnace ventilation systems or had been involved in fire suppression of the smoldering pipes, who were interviewed, claim they had not had fire extinguisher training the whole time they have been in maintenance."[60]

NFPA states that "Nearly all vaporizing liquid-fire extinguishing agents react violently with burning aluminum, usually serving to greatly intensify the fire and sometimes resulting in explosion."[61] "Class B extinguishing agents will usually greatly accelerate combustible aluminum dust fires and can cause the burning metal to explode."[62] The material safety data sheet (MSDS) for the aluminum ingots used by Hayes Lemmerz clearly states "In case of aluminum fire, use a class D dry-powder extinguisher. Do not use water...."[63]

The Class D fire extinguisher is the appropriate extinguisher to use on an aluminum fire. A high level manager at the plant acknowledged his awareness that ABC type extinguishers are not to be used for metal fires. Despite this, "ABC type fire extinguishers were available for use in the area when the major fire hazard was the aluminum dust and chips. Class D extinguishers were also available in the area."[64]

Following a magnesium dust fire that occurred at the Hayes Lemmerz plant on 3/10/00, a supervisor notes in his fire incident report, "We only have 1 Class D fire extinguisher that is charged. These need to be filled today. We need to purchase more of the Class D extinguishers...."[65]

"Welding was performed in an area where combustible aluminum dust and chips were present and ... to close holes on the cyclones and piping when there was combustible aluminum dust inside of them."[66] Hayes Lemmerz had knowledge of, and referenced NFPA 51B in their own procedures. NFPA 51 B cautions that welding should "not be permitted ... in the presence of explosive atmospheres ... in areas with an accumulation of combustible dusts that could develop explosive atmospheres."[67]

Between mid-September and the October 29th explosion, it appears there were at least six work orders to repair holes and leaks in the chip melt system.[68]

One employee states, "About one week before the explosion, I had to weld a patch on the tube for cyclone above the well. We had a welding permit which was issued by_____There was chip and dust all around area. Did not clean inside the pipe and there was (sic) a few holes burnt into the pipe. I had fire extinguisher training about 7 to 8 years ago. Sometime (sic) when we had pipe fires we would disassemble some to clean it out. We have no fire retardant clothing. We have had several fires over the years and it was always in the piping. And there was always lots of dust and chips everywhere."[69]

"I was told ... to go back inspect and repair cyclone on number 5 reverb. I had _____ as a fire watch. After inspecting the cyclone, I noticed that above and below the flange that there were a number of small holes ranging in diameter. I then got an arc welder and repaired two of the holes ... and then stopped (.) The metal was so thin I couldn't weld on it without causing the holes to expand. I notified _____ that the metal was too thin and it needed (sic) replaced."[70]

IOSHA determined that Hayes Lemmerz had permitted cutting, welding "in the presence of explosive atmospheres (mixture of flammable gases, vapors, liquids, or dusts with air), or explosive atmospheres that could develop inside uncleaned or improperly prepared tanks or equipment which had previously contained such materials, or that could develop in areas with an accumulation of combustible dusts."[71]

Finally, "The electrical installations in the number 5 furnace area ... breaker panel, disconnect boxes, control panels, and junction boxes are not approved for a Class II Division 1 location."[72]

A Class II, Division 1 location is defined as a location "(a) in which combustible dust is or may be in suspension in the air under normal operating conditions, in quantities sufficient to produce explosive or ignitable mixtures; ... or (c) in which combustible dusts of an electrically conductive nature may be present."[73]

Astonishingly, a high level manager at the plant for over three years at the time of the explosion, tells IOSHA that he was aware that pipes were worn, eroded and leaking. But he also states, "I was not aware of explosive potential of aluminum dust. I understand that the dust would smolder and I understood that it was combustible dust …". As noted by the IOSHA inspector, "How then could he ensure the hazardous nature of aluminum dust could be transmitted to employees so that they understood the significant nature of the material they were dealing with."[74] In the IOSHA inspection in 1997 cited previously, Hayes Lemmerz is told "The employer shall apprise employees of the fire hazards of the materials and processes to which they are exposed."[75]

A high level manager at the plant for one year at the time of the explosion acknowledges that the fire issue would come up in production meetings, but he had never seen the flashes in the chip well that the employees talked about, "only small flames when the chips were introduced." On the housekeeping conditions, he states, "I noticed some general housekeeping opportunity with respect to chips on the floor" and he thought he most likely brought it to the attention of the (supervisor). He also acknowledges that he is "aware that ABC type extinguishers were not to be used for metal…."[76]

Tammy Miser posits, "as a worker, you just accept that your employer is going to protect you."

"We … can take legal action for hot coffee, unkind words, harming ourselves (smoking) and injuring or killing animals. When it comes to human lives, the system pushes the papers and shrugs it off to the point that these things … outweigh a death."[77]

The circumstances of the death of her brother have made Tammy Miser an activist. Miser, together with a colleague who also suffered a family workplace death, started United Support and Memorial For Workplace Fatalities (USMWF). The organization's purpose is to be a source of "support for individuals, companies and other organizations dealing with the repercussions of a workers death and the prevention thereof." Since its inception, USMWF has been recognized and honored nationally.[78]

Miser has worked with other groups to get Congress to amend the federal Occupational Safety and Health Act to change the provisions regarding citations, to increase the criminal penalties for reckless disregard for safety that results in worker deaths, and to pass legislation to require OSHA to regulate combustible industrial dusts.

On April 29, 2004, IOSHA entered into a settlement agreement with Hayes Lemmerz. Hayes Lemmerz agreed to suspend operation of "any and all pneumatic conveyance systems for dry aluminum chips and fines as well as the rotary kiln dryer at the Huntington facility." It was understood that the company had suspended similar operations at its other facilities throughout the United States until it could "determine if there is a safer method for handling aluminum chips for reprocessing." Hayes Lemmerz also agreed to purchase "two additional rolling class D fire extinguishers" to add to the 1 existing rolling class D fire extinguisher; to extend the access road around the facility, thus providing better access for responding emergency vehicles; to provide training on hazards associated with explosive dusts and fire-fighting protocols for various management employees who will in turn train employees; to report all facility fires to the Huntington Fire Department and to invite the "Fire Department to tour its facility at least annually;" to make a $10,000. donation to the St. Joseph's Burn Unit, and to pay a $42,000 fine.[79]

One man is dead, one man has been burned over 40% of his body, 5 others have been injured, and many more were put at risk. Hayes Lemmerz, a multi-billion dollar international company was fined the sum of $42,000.

By contrast, a radio station in Miami has been fined $55,000 by FCC for "broadcasting sexually explicit material."[80] One must continuously ask where are our priorities?

For the fiscal year in which Shawn Duane Boone was killed, the company reported a $62 million operating profit. Revenues for 2003 were $2.1 billion.[81]

The citations and fines appear so inadequate, when weighed against the death of one worker, the injuries and the exposure of so many to burns, smoke inhalation, and death. A partial reason for this is the lack of specific OSHA regulations governing aluminum oxide and other dusts in the industrial setting; a fatal oversight.

In 2003, the year of Shawn's death, 14 workers in the United States died from combustible powder explosions.[82]

The State of Indiana's IOSHA investigation was extensive and comprehensive. But the limits of the regulations and the pressure to negotiate downward on citations and fines in order to move swiftly to closure, in the interests of protecting the remaining workers at the plant in the short term, works against holding the corporation accountable, which would create far safer conditions for workers in the long run. As was noted in this investigation, several of the citations were a repeat from previous OSHA investigations at the same plant.

One life has been lost, others have been injured, families are left suffering the pain, grief and loss. The settlement agreement contains the standard protection from harm for the employer, i.e., that "nothing contained in this agreement ... shall be interpreted to demonstrate causation of the events on October 29, 2004, or the resulting injuries to any individual."[83]

The US Chemical Safety and Hazard Investigation Board investigators found that "the company did not address why the chip drying system was releasing excess dust, and did not identify or address the dangers of aluminum dust ignition, despite having a history of small dust fires inside the factory." It also noted that "Hayes Lemmerz did not ensure the dust collector system it ordered was designed in accordance with guidance in ... fire code published by the National Fire Protection Association." CSB Chairperson Carolyn Merritt stated that "aluminum dust is among the most explosive of all metal dusts and the conditions in dust collectors that are not properly designed, installed or maintained present the ideal environment for an explosion and fire.

"This accident followed a classic syndrome we call 'normalization of deviation,' in which organizations come to accept as 'normal' fires, leaks, or so-called small explosions. The company failed to investigate the smaller fires as abnormal situations needing correction or as warnings of potentially larger more destructive events. The CSB almost always finds that this behavior precedes a tragedy."[84]

"It is so much easier for them to pay a fine or get a warning, then it is to fix the problem and save a life," states Tammy Miser, Shawn's sister.[85]

And what was Hayes Lemmerz responsibility under workers' compensation? Indiana Workers Compensation Board reportedly sets payment to surviving spouses and dependents at a percentage of the worker's weekly salary for 500 weeks, not to exceed a combined total of $294,000, or a maximum of $588 per week.[86]

But Shawn Boone was neither married, nor had children. There were no dependents. It appears that Hayes Lemmerz had no workers' compensation responsibility beyond medical, medical transport and, possibly, minimal funeral expenses. The latter varies by state.

Historically, workers' compensation has limited compensation to surviving spouse and/or dependents. The pain and suffering of the family, the loss of their beloved brother, son, uncle in this excruciatingly painful way is not recognized in the workers' compensation system.

In early 2006, Hayes Lemmerz International is reported to have announced its intent to close the Huntington plant later that fiscal year.[87]

References:

1. Indiana Department of Labor, Occupational Safety and Health Administration (IOSHA), Inspection No. 123987323, 6/06/96.

2. CSB Investigators Find Likely Source of Dust Explosion at Indiana Automotive Plant. United States Chemical Safety and Hazard Investigation Board, 11/5/03, www.csb.gov.

3. IOSHA, Inspection No. 306860461, Coverage Information, p. 6, Citation 01, Item No. 004, p.2

4. ibid.

5. Investigation Report Aluminum Dust Explosion Hayes Lemmerz International-Huntington, Inc., Report No. 2004–01-I-IN, U.S.Chemical Safety and Hazard Investigation Board, September, 2005, p.19.

6. Plant blast investigation begins. Associated Press, October 30, 2003, Indy Star, www.indystar.com/print/articles/0/088099–1260–102.html.

7. IOSHA, Inspection No. 306860461, Coverage Information, p. 8.

8. IOSHA, Inspection No. 306860461, handwritten notes.

9. Office of Fire Marshal, Report of Fire Investigation, Case # 35W223, Huntington, Indiana.

10. Allen County Coroner's Report, Case No. 03C000448, dated 11/06/03.

11. Alcan Aluminum Metal Material Safety Data Sheet (MSDS), revision date 1993/04/16, cited in IOSHA, Inspection No. 306860461, Citation 01, Item No.001, p. 6.

12. Recommendations for storage and handling of aluminum powders and paste, TR2, Background: Why Powder Explodes. The Aluminum Association, Inc., Arlington,VA, February 2006.

13. IOSHA, Inspection No. 306860461,handwritten notes, Statement 8A.

14. IOSHA, Inspection No. 306860461, Coverage Information, p. 6.

15. ibid.

16. IOSHA, Inspection No. 306860461, Citation 01, Item 001, p.2.

17. IOSHA, Inspection No. 306860461, Citation 01, Item 004, p.4 and hand-written notes.

18. ibid.

19. IOSHA, Inspection No. 306860461, Coverage Information, p. 6.

20. IOSHA, Inspection No. 306860461, handwritten notes.

21. IOSHA, Inspection No. 306860461, Coverage Information, p. 7.

22. ibid., Statement 7B.

23. IOSHA, Inspection No. 306860461, Coverage Information, p. 5 and hand-written notes.

24. IOSHA, Inspection No. 306860461, Coverage Information, p. 7.

25. ibid.

26. Hayes Lemmerz, Huntington plant, Environmental Emergency Response Plan, 11.0 and 11.3.4 in IOSHA Inspection No. 306860461.

27. IOSHA, Inspection No. 306860461, Citation 01, Item No. 004, p.4 and Statement 2A, p.2.

28. IOSHA, Inspection No. 306860461, Coverage Information, p. 9.

29. IOSHA, Inspection No. 306860461, Citation 01, Item No. 004, p.4; and Statements 1A, p.2; 5A, p.2; and 3A, p.2.

30. IOSHA, Inspection No. 306860461, Coverage Information, p.9 and Citation 01, Item No. 004, p.4.

31. IOSHA, Inspection No. 306860461, Citation 01, Item No. 002, p.3 and Citation 01, Item No. 004, p.4.

32. IOSHA, Inspection No. 306860461, Citation 01, Item No. 002, p.3 and Contractor's worksheet for 10/14–10/23/03.

33. IOSHA, Inspection No. 306860461, Coverage Information, pp.9,10; and Statement 7, pp. 1–3.

34. IOSHA, Inspection No. 306860461, Coverage Information, p. 10; and Statement 8.

35. IOSHA, Inspection No. 306860461, Coverage Information, p. 10; and Statement 9A.

36. IOSHA, Inspection No. 306860461, Citation 01, Item No. 002, p.5.

37. IOSHA, Inspection No. 306860461, Coverage Information, p. 10.

38. Office of the Fire Marshal, Report of Fire Investigation, Number 35W223, Huntington County, Indiana.

39. IOSHA Inspection No. 306860461, Citation No. 01, Item No. 001, p.2; and IOSHA Air Sampling Report, Sampling #913426672, dated 10/31/03.

40. IOSHA, Inspection No. 306860461, Coverage Information, p. 10.

41. National Fire Protection Association Codes (NFPA) 484, Standard for Combustible Metals, Metal Powders, and Metal Dusts, Section 4.4.4., 2002 Edition.

42. IOSHA, Inspection No. 306860461, Citation 01, Item No. 001, pp.1,3,5.

43. National Fire Protection Association, NFPA 484, Standard for Combustible Metals, Metal Powders, and Metal Dusts, Annex A.4.4.3, 2002 Edition.

44. Investigation Report Aluminum Dust Explosion Hayes Lemmerz International-Huntington, Inc., Report No. 2004–01-I-IN, U.S.Chemical Safety and Hazard Investigation Board, September, 2005, p.73.

45. IOSHA, Inspection No. 306860461, Coverage Information, p. 11.

46. IOSHA, Inspection No. 306860461, Citation 01, Item No. 006a, p.3.

47. IOSHA, Inspection No. 306860461, Citation 01, Item No. 001, p.7, and handwritten statement.

48. Cleaning Sign-off Sheets, in IOSHA Inspection No. 306860461.

49. IOSHA, Inspection No. 125377697, dated 4/18/97, cited in IOSHA Inspection No. 306860461, Citation 01, Item No. 004, p.7.

50. IOSHA, Inspection No. 306860461, Citation 01, Item No. 004, p.7.

51. IOSHA, Inspection No. 306860461, Coverage Information, p. 11.

52. ibid.

53. National Fire Protection Association, NFPA 484, Standard for Combustible Metals, Metal Powders, and Metal Dusts, 4.6.2.1, 4.6.2.3, 2002 Edition, cited in IOSHA, Inspection No. 306860461, Citation 01, Item No. 005, p.3.

54. IOSHA, Inspection No. 306860461, Citation 01, Item No. 005, p.8.

55. IOSHA, Inspection No. 125377697, dated 4/18/97, citing 1910.156(e) (1) (i),cited in IOSHA Inspection No. 306860461, Citation 01, Item No.005, p.6.

56. Letter Barnes & Thornburg re Hayes Lemmerz International, to IOSHA, p.2, February 24, 2004, in IOSHA Inspection No. 306860461.

57. National Fire Protection Association, NFPA 484, Standard for Combustible Metals, Metal Powders, and Metal Dusts, 4.5.5.1, Fire Fighting Organization, 2002 Edition and IOSHA Inspection No. 306860461, Citation 01, Item No.001, p.l.

58. (Company performing air emissions testing) Statement on the Explosions at Hayes Lemmerz International, Huntington, IN, dated 10/29/03, cited in IOSHA Inspection No. 306860461.

59. National Fire Protection Association, NFPA 484, Standard for Combustible Metals, Metal Powders, and Metal Dusts, Fire fighting Organization, 4.5.5.1.1, 2002 Edition, cited in IOSHA, Inspection No. 306860461, Coverage Information, p. 11.

60. IOSHA, Inspection No. 306860461, Coverage Information, p. 11.

61. National Fire Protection Association, NFPA 484, Standard for Combustible Metals, Metal Powders, and Metal Dusts, 4.5.2.1, 2002 Edition.

62. National Fire Protection Association, NFPA 484, Standard for Combustible Metals, Metal Powders, and Metal Dusts, 4.5.2.6.3, 2002 Edition.

63. Alcan Aluminum Metal Material Safety Data Sheet (MSDS), revision date 1993/04/16, Section 5, in IOSHA, Inspection No. 306860461.

64. IOSHA, Inspection No. 306860461, Citation 01, Item No. 002, p.6 and Coverage Information, p.11.

65. Hayes-Lemmerz Fire Incident Report, dated 3/10/00, 4:15 am. and IOSHA, Inspection No. 306860461, Citation 01, Item No. 002, p.6.

66. IOSHA, Inspection No. 306860461, Coverage Information, p. 12.

67. National Fire Protection Association, NFPA 51B, Standard for Fire Prevention During Welding, Cutting and Other Hot Work, 5.2 (3),(5), 2003 Edition and IOSHA Inspection No.306860461, Citation 01, Item 006(a).

68. Letter Barnes & Thornburg, re Hayes Lemmerz International, to IOSHA, dated 12/18/03, in IOSHA Inspection No. 306860461.

69. IOSHA, Inspection No. 306860461, Citation 01, Item No. 006a, p.3, Statement 4.

70. IOSHA, Inspection No. 306860461, Citation 01, Item No. 006a, p.3, Statement 3B.

71. IOSHA, Inspection No. 306860461, Citation 01, Item No. 006b, p 1.

72. IOSHA, Inspection No. 306860461, Citation 01, Item No. 002, p.2.

73. 29 CFR, 1910.399, Class II locations, (i) Class II, Division 1.

74. IOSHA, Inspection No. 306860461, Citation 01, Item No. 003, p.2, Statement 13, other.

75. IOSHA, Inspection No. 125377697, dated 4/18/97, cited in IOSHA, Inspection No. 306860461, Citation 01, Item No. 003, pp.4,5.

76. IOSHA, Inspection No. 306860461, Citation 01, Item No. 005, p.5, and Citation 01, Item No.001, p.4.

77. Letter Tammy Miser to State Senator Tom Weatherwax, dated 11/21/03.

78. United Support and Memorial For Workplace Fatalities, www.usmwf.org/Mission.htm.

79. Settlement Agreement, IOSHA and Hayes Lemmerz International,Huntington,Inc. dated 4/29/04 .

80. Clifton, Alexandra Navarro. FCC Fines WQAM $55,000. South Florida Sun-Sentinel, p.3D, 11/24/04.

81. Hayes Lemmerz News Release, dated April 6, 2004, Hayes Lemmerz International Reports Strong Operating Profit for Fiscal Year Ended January 31., www.hayes-lemmerz. com/press_kit/html/20040406_q4_earnings .html and Hayes Lemmerz Fact Sheet at www.hayeslemmerz.com/press_kit/html/hayes_lemmerz_fact _sheet.html.

82. CSB Investigation Information Page. Combustible Dust Hazard Investigation. United States Chemical Safety and Hazard Investigation Board, October 1, 2004, www.csb.gov/index.cfm?folder=current_investigations&page=info&INV_ID=53.

83. Settlement Agreement, IOSHA and Hayes Lemmerz International, Huntington, dated 4/29/04, p.6.

84. U.S. Chemical Safety and Hazard Investigation Board. CSB News Release. CSB Determines Fatal 2003 Incident at Hayes Lemmerz Plant in Indiana Most Likely Caused by Explosion in Dust Collection System; Company Did Not Identify or Control Hazards of Aluminum Dust.10/5/2005. www.csb.gov/index.cfm?folder=news_release&page=news&NEWS_ID=243.Note: The CSB is a federal agency. It does not issue citations or fines. It does make safety recommendations to industry organizations, labor groups, and regulatory agencies.

85. Letter Tammy Miser to State Senator Tom Weatherwax, dated 11/21/03.

86. Green, Rebecca S. A year later, grief is replaced by activism. The Journal Gazette, 10/31/04, at www.fortwayne.com/mld/journalgazette/news/10063242.htm

87. Hayes Lemmerz to Close Huntington Plant. Inside Indiana Business, 3/17/06, www.insideindianabusiness.com/newsitem.asp?ID=17235.

CHAPTER 6

▼

CONSTRUCTION: A MOST DANGEROUS INDUSTRY

The construction industry has consistently had the largest number of workplace fatalities recorded by U.S. Bureau of Labor Statistics(BLS). Twenty percent (20%) of all workplace fatalities in the US recorded by BLS, 1,221 for 2002, were in the construction industry.[1]

The recorded occupational injury and illness rate for construction workers is the second highest for any industrial sector in the nation, at 7.1 per 100 workers for 2002.[2]

*   *   *   *

Charles Wayne Wiggins was a pipe fitter. He worked at oil refineries, replacing the old pipes and the fittings, and welding them out, while the refinery remained in shutdown, until the work would be finished. Wayne would travel to each job in his truck, often for long distances, all across the U.S. This particular facility was in Illinois.

Once there, the workers brought in during shutdown to do the repairs would live in what Wayne referred to as the "man camp," temporary housing and a commissary to provide meals for the workers.

Wayne Wiggins had been a pipe fitter for 25 years—until March 7, 2003. That was the day he fell and injured his shoulder.

The stairs were icy from the cold weather, and greasy from the oil at the refinery. "Both feet went out from under me," Wayne remembers. He grabbed the stair railing with his left hand, and hit his back and shoulder on the way down. Wayne first felt the pain in his back. He was taken to the local health care center. The initial concern was his back, but that felt fine after a few physical therapy treatments. His left shoulder is where he continues to have the problem. He had injured the rotator cuff, and the surrounding muscles and tendons. Wayne wasn't sure of the technical terms, but he thought he remembered tendonitis being one of the terms used. "They made a pretzel out of it," he summarizes.

The rotator cuff is made up of muscles and tendons that connect the upper arm bone with the shoulder blade. It also holds the ball of the upper arm bone in the shoulder socket. The shoulder is designed specifically for movement.[3] One of four tendons connects to each of four muscles that moves the shoulder in a certain way.[4] And it is this constellation of tendons and stabilizing muscles that Wayne damaged when he fell.

For the next two weeks at work, he was sent to the "dress out" trailer, until he was let go on March 20th. By having Wayne show up for work, even though he wasn't physically able to work, his employer could exclude Wiggins' injury from the employer's count of injuries that require days away from work, and would thus impact the employer's workers' compensation insurance premium.

It's very difficult mentally to sit around and do nothing for 12 hours a day, Wayne explains. After a few days, he asked people whether there was anything he could do, e.g. could he get screws, bolts for people, etc. Wayne used the elevator to deliver the small items he could carry, since he couldn't use the stairs.

Wiggins says the contractor's director of safety told him that there had been over 100 injuries during the thirty-day shutdown on the same oil refinery where he was injured. But despite this obvious site safety red flag, he also hinted to Wayne that he might be faking it, trying to milk the system. As Wayne says, "in 25 years as a pipe fitter, I have never had a workers' compensation claim until now."

Wayne Wiggins' physician told him he needed an operation to fix the shoulder. The workers' compensation insurer required a second opinion by a physician of the insurer's choosing. The designated workers' compensation physician saw him for one brief visit and rendered the opinion that Wayne didn't need further medical treatment. He curtly told Wayne he wasn't there to answer any of his questions. This physician also proceeded to lose the x-rays and the MRI that Wayne had brought to him, which Wayne had to pay to replace.

Once the workers' compensation physician had made his determination, not only was the operation denied, but the check for 80% of wage compensation, or $1004 per week, stopped.

It was the latter that put Wayne and his wife into serious financial jeopardy. Bills came due with no money to pay them. Wayne, who had had excellent credit, now has filed for bankruptcy, his wife has lost her car, and they are just holding on. When he worked, he made $36/hour. Now, a disability pension from the U.S. Marine Corps of $316/month is their major source of income, supplemented by a little money that his wife makes.

Wayne had the needed operation in April 2004. He had signed an agreement that the cost of his operation would be reimbursed to his insurance when the workers' compensation case was settled. Recently he got a call from the hospital saying that his insurance isn't going to cover it, and that he owes the hospital the full cost of the surgery.

Wayne also had coverage that would cover his truck payments in case of disability. Now that insurance is questioning whether they will honor it. Wayne believes the difficulty is because of the two conflicting medical opinions on his condition. He had forgotten about the coverage and had paid his truck payments as long as he could. He now says that at the very least they ought to return the payments made from March 2003 through October 2003, the point at which the "quack" opinion, as he calls it, was made.

Wayne Wiggins is angry, but with no way out of his plight, it turns to depression, for which he takes medication. He used to enjoy fishing, but now he can't even do that because of the shoulder.

At 56 years old, Wayne is in pain, unable to work, in financial ruin. This is what the workers' compensation insurance industry phrase, "Starving them out," means. Wayne Wiggins, like millions of workers before him, has been "starved out" by workers' compensation.

"This is not the way it's supposed to be. If I injured someone, I would be held accountable. Why aren't the insurance companies held accountable?" Wayne asks.

"We need a March on Washington. We need people to listen. We need to change this," he adds.

*   *   *   *

As Tom Matthews, Business Manager and Secretary-Treasurer for Laborers International Union of North America, Construction & General Laborers Local Union, No. 767 in West Palm Beach, Florida sees it, "One of the largest problems we have had in the construction industry was the independent contractor problem, which has been fueled by workers' compensation and the cost of it." When Matthews first began to explore this issue, he was hearing from long-established owners that they were having a difficult time getting jobs. "Too many people were beating them out of work. That started me thinking, why is it that somebody who has been in business, at the time for 25 years or more, is having a hard time getting work—he's so established. I represent the laborers that work for this contractor, other contractors. The more jobs they get, the more jobs for my people." He started to see that a lot of people were being paid under the table. Matthews started looking at workers' tax forms, and that's when he discovered the 1099s (the form filed by an independent contractor).

Occupations within the construction industry are very high-risk occupations. An employer's workers' compensation insurance premium is based on the size of the payroll, a percentage rate based on the classification of the specific occupation, e.g. ironworkers, and a modification according to the specific employer's risk or claim experience, Matthews explained. Some construction industry employers, in an effort to reduce their cost of workers' compensation insurance, were hiring workers, which they then designated as "independent contractors." In this way, these workers were not included in the employer's workers' comp premium, nor would social security, taxes, or unemployment insurance be deducted. Since workers' compensation insurance premiums can range from 20% to 100% of payroll, all this gives the unscrupulous contractor the ability to undercut long-established honest contractors on bids—by as much as 35%, says Tom Matthews. "So what had been a snowball turned into an avalanche, not just in Florida, but a lot of places in the country."

This was when he and the Coalition for Independent Contractor Reform that he started through the Laborers' International Union decided to take action, explained Tom. Matthews points out that the Coalition included non-union contractors as well. They too wanted to "get a level playing field." The group walked in on the State Insurance Commissioner's hearing with the National Compensation Insurance Carriers, who were requesting a rate hike. "We filled that place up with employers who were being screwed." He gives the example of one iron-working business owner, who pointed out that she had paid $26,000. in workers' compensation premiums the prior year, while her competitor with the same number of employees only paid $3,000. She had hired an investigator to find this out and had brought the evidence to prove her claim.

Tom Matthews and the Coalition were instrumental in bringing about grand jury hearings on workers' compensation that uncovered the premium fraud by companies, the looking the other way of both insurance carriers who sold "minimal insurance policies," and state investigators, who could see 30 workers on a site when the insurance policy stated three employees.

Workers' Compensation insurance was a $2.1 billion industry in Florida in 1996. What the Grand Jury uncovered when they conducted their investigation was that fully 13% of their investigation sample of 23,000 employers were operating without any workers' compensation coverage, and a repeat study the following year, with a larger sample, showed an increase to 13.6% of employers in the state operating without any workers' comp coverage.[5] The report identified a major reason for this: That some employers caught underinsuring were prosecuted criminally and/or fined, while those who purchased no insurance got away with a slap on the wrist, if that. For example, a Tampa-based business that was cited for being without insurance in 1994 was fined $1000, was cited again in 1996, and three times in 1997. Despite this, no fines were ever collected and the company continued to operate without interruption, nor was any stop-work order ever issued. The state's own Department of Labor (DOL), Division of Workers' Compensation had created an incentive for employers, especially in the high risk construction industry, to purchase no insurance.

Some construction industry employers were also fraudulently misclassifying employees, e.g., a roofer classified as a clerical employee, for a cost savings of approximately $79. per $100. of payroll ($.89/$100. for clerical vs. $80./$100 for roofer). And, as mentioned previously, the fraudulent labeling of employees as independent contractors was being widely used. However, DOL was counting all these underreporting and misclassifying businesses as being in compliance with statutes—a determination that defied logic, according to the Grand Jury. DOL's primary motivation seemed to be to avoid becoming a burden to businesses operating in the state, even if that reduced revenues to the system by millions of dollars, and allowed unscrupulous businesses to fraudulently gain "an unfair economic advantage over the honest business that is paying into the system."[6]

Insurance companies also were not making any effort to uncover fraudulent practices. Indeed they were selling so-called "minimum premium policies" to employers who report no employees on payroll at time of purchase.[7] At the same time, they would come to the State Insurance Commission and request higher rates.

The end result of this gross employer fraud is that Florida has one of the highest workers' compensation premium rates in the nation at the same time that benefits are among the lowest in the nation.

*   *   *   *

It was after 6:00 PM on July 22, 2004, in Hobe Sound, Florida, when one of the buildings being constructed for a luxury townhouse condominium complex, ironically called "Tranquility," collapsed.

Carpenters Gabriel Suarez, 31, and Leonardo Navarro, 23,[8] were standing on the first floor in Building No. 9. Having completed their work for the day, they were waiting for their ride home. On the second and third floors, concrete was being poured onto tunnel forms of the second floor walls and third floor deck, "when the shoring system supporting the building failed and the building collapsed."[9]

Gabriel Suarez and Leonardo Navarro were trapped and buried under the collapsed structure and debris. Fire rescue, construction workers, and two passersby attempted to assist. Because of the unstable condition where the shoring and concrete had fallen, the two men were unable to be removed until early the next morning, at which point the mortuary was called.[10] Gabriel Suarez and Leonardo Navarro were dead from "traumatic asphyxiation with crush injuries." They had had the life crushed out of them, painfully, agonizingly.[11]

Antonio Perez, 31,[12] was pinned to the ground by large falling pieces of concrete. Fire Rescue and others assisted in freeing Perez, after which he was transported to the hospital in critical condition.[13] His pelvis was crushed, his elbow broken, but he had survived.

In all, five workers were injured. One co-worker described hearing the screams for help as he crawled out through the rubble. Six others escaped injury, including a thirteen-year-old boy who was reportedly working at the site. Florida law prohibits anyone under 16-years from working on a construction site.[14]

What Mr. Perez, the other injured workers, and the families of Messrs. Suarez and Navarro soon discovered was that their employer, Mac's Construction & Concrete, Inc. did not have workers' compensation insurance to cover any of the construction workers at the site. No matter how long they had worked for Mac's Construction—reportedly six years in Mr. Perez' case, they weren't covered. As it would turn out, there was no evidence that the ten-year old Mac's Construction

& Concrete, Inc. had carried Workers' Compensation insurance on any of its construction workers since March 1997.[15]

Over the ensuing months, the Perez family saw their debts mount and their savings depleted, while Mr. Perez was hospitalized for several weeks, underwent three operations, and remained virtually immobile for months afterward.[16] It appears that his impairments are permanent.[17]

OSHA began its investigation finding that it was only two days earlier, on July 20, 2004, that concrete work had begun on Building No. 9 (the site of the incident) using the tunnel forms.... "the tunnel forms were placed in their final location to begin casting concrete over the second floor and in the walls between the first and second floor. Concrete began to be placed on the forms at about 4:00 PM and was completed by about 8:00 PM."[18]

The next day the tunnel forms were loosened, and moved from the south end to the north end of the building, and again concreting took place on the second floor and walls.

The following day, July 22, 2004, the same routine "was followed to cast the third floor and the wall between the second and third floors.... tunnel forms at the north part were loosened and were rolled out and placed on the south part on the second floor," after which concreting began at about 3:00 PM.[19]

According to one long-time licensed building contractor, it takes seven to twenty eight days for concrete to set properly. By pouring the third floor concrete just a few days after the concrete on the second floor, it appears it never had time to properly cure.[20]

OSHA's investigation revealed that the "single post shoring and reshoring support system was not capable of carrying the imposed load." Management at Mac's Construction & Concrete had "failed to insure the adequately sized post shores and form jacks were correctly installed prior to allowing the concrete pour to begin." Fully half of the tunnel formwork jacks "were not screwed down to the slab in the area where the building failed," and "single post shores that were placed under the walls were spaced at intervals that subjected them to loads beyond their capacity." Intervals of four feet were used instead of the maximum allowable three feet six inches for the type of shore used. Mac's management officials had had ample warning of the problem "when post shores were observed deformed and bent under load during the concrete pour at building No. 2" in the same complex.[21]

Apparently, the Tranquility complex had originally been designed for masonry block construction. Earlier in 2004, Mac's had suggested to the contractor, Allied Capital and Development, the change to tunnel formwork in order to streamline the process and cut costs.[22] In 2006, the manufacturer of the tunnel forms advertised a 25% savings in time and a 15% savings in cost associated with the use of their tunnel forms, and a "completed shell months sooner than standard forming methods."[23]

The late re-design was accepted, and it was also agreed that Mac's, as the concrete contractor, was responsible for the submission of the shoring and reshoring plans. However, no formwork support plans were provided, in violation of OSHA regulations and state building code. Florida Building Code (FBC) 1906.2.2.1 states, that "before starting construction, the contractor shall develop a procedure and schedule for removal of shores and installation of reshores and for calculating the loads transferred to the structure during the process."[24] Mac's apparently never provided the detailed shoring plan for the site that should have been submitted for approval to the structural engineer for the project.[25]

Professional Engineering & Inspection, Inc.(PEICO) had been hired to perform inspections and concrete sampling and testing for strength. On June 28, 2004, PEICO refused to allow "the initial concrete pour into tunnel formwork because there was no single post shoring plan available at the worksite."[26]

A hastily hand-written one pager was put together. It failed to specify any building numbers to which it was applicable, nor "the size and strength of the single post shores to use; it did not contain any sketches to illustrate the location of shores; it did not address the size location and spacing of shores under the walls which began from the second floor up; and finally, it did not specify the required concrete strength at which time the tunnel forms could be removed."[27] The American National Standard Institute(ANSI), in its standard, states, "The shoring and reshoring drawing and/or specifications shall include details of unusual conditions such as heavy beams, sloping areas, raps, and cantilever slabs, as well as plan and elevation views, minimum values of concrete strength for formwork removal, and maximum allowable loads for which the formwork system is designed."[28] PEICO allowed the concrete pour to proceed.[29]

Only a week or two earlier, at building #2, several single post shores buckled under load. Mac's Construction installed additional single post shores next to the failing ones, in order to provide additional support. But Mac's didn't investigate the cause of the failure, nor did it consult the structural engineer to identify the cause of the failure. Mac's own foreman had told Mac's that the post shores "appeared to be too thin for the application." In fact, there was a disconnect between Mac's Construction and the structural engineer on the type of post shoring used, with the type of shoring actually used being of a lower load bearing capacity than what the project's structural engineer thought was going to be used.[30]

According to OSHA' s expert structural engineer, there are only two reasons that post shores fail: either "inadequate strength or improper spacing of posts." It would turn out that both failures were present at Tranquility. The failure on building #2 should have been a warning.[31]

OSHA noted that no inspection of the placement of post shoring was performed prior to, during, or after the concrete placement.[32] At building #9, a day or two before the disaster, "the tunnel forms were removed prior to attaining 30% of the compression strength specified by the project's structural engineer," although it appears instructions were not written, but given verbally to Mac's Construction.[33]

Formwork was being removed at an average compressive strength of 27.8%. Laboratory analysis on sections of building #9 showed a compressive strength ranging from 979 psi (pounds per square inch) down to 687 psi, with an average compressive strength of 833 psi, when it should have been at 900 psi before forms could be removed. "The lower strength of the concrete was believed to have contributed to the collapse."[34]

As stated in a July 8, 2004 letter from Allied Capital & Development, there had been "numerous referenced conversations, admonishments, and verbal warnings about serious, hazardous and unsafe conditions.." including lack of perimeter fall protection, for which Mac's Construction had been cited many times in the past on other construction sites.[35]

In February, 2003 on a construction site in Orlando, Florida, Mac's Construction and Concrete, Inc. exposed several workers to risk of falls, ranging from 9.5 feet to over 19 feet. A 19 foot unprotected fall can result in death. Workers "were not protected from falling by use of guardrail systems, safety net systems or personal fall arrest systems. Lack of knowledge of risk was not a factor, OSHA noted. The exposure was "visually apparent" and the "employer was in the area of exposure."[36] Mac's Construction was fined less than $2000.

Less than seven months later, Mac's was cited for the same violation, this time at a site in Dania Beach, Florida.[37] In all, OSHA had cited Mac's for no less than 21 violations on nine sites over the preceding 27 months, many of which could be lethal to workers.[38] This time, in addition to the previous violations, Mac's had failed to erect, support, brace, and maintain formwork, "so that it would be capable of supporting without failure all vertical and lateral loads that could reasonably be anticipated...."[39]

Earlier on the day of the accident, July 22nd, Allied Capital & Development wrote an urgent letter to Mac's Construction, requesting that the shoring plan be submitted to the structural engineer of record for approval immediately.[40]

In buildings #2 and #10, OSHA inspectors noted and photographed bent shoring, shoring loaded to stress levels and the use of mason blocks to support single post shoring. The latter was also observable in building #9. OSHA's national office structural analysis would note that "the concrete contractor did not provide an adequate number of single post shores on the first floor to support the weights of the walls of the second and third floors and the weight of the second and third floor slabs. The posts buckled under the load and precipitated the failure (collapse) of the building…. A number of tunnel formworks were not properly placed because their leveling jacks were not turned down to the floor slab to properly transfer the loads." In fact, approximately half of the "outer tunnel formwork jacks were not screwed down to the slab in the area where the building failed." Another factor was the spacing between the shores, which ultimately imposed loads exceeding the shore's capacity, causing it to collapse.[41]

PEICO, the inspection firm, was hired by the developer to "act as a substitute for a Martin County requirement that a county inspector verify compliance with the Florida Building Code."[42] Were there then no inspections by the Martin County Building Department? A state law quietly passed in 2002 reportedly allowed developers to hire a consultant to do the inspections normally expected of local government building departments.[43] State of Florida thus allowed a private vendor to be both paid by a project and an inspector for the same project.

The contractor, the inspection company, and Mac's Construction; all were aware of serious flaws and safety hazards.[44] The work continued nonetheless—with tragic results for Gabriel Suarez, Leonardo Navarro, Antonio Perez and their families.

Gabriel Suarez left a wife and two daughters, aged 7 years and 5 years of age. His youngest daughter cries a lot because she misses her father and, too young to understand, tells her mother that she wants to die too, so she can be with her Dad. Suarez had plans to move his family into a condo back home by Christmas.[45]

Leonardo Navarro leaves a wife and a two-year old daughter. He had helped support his parents, and paid for his sisters' and brothers' educations. Navarro had dreams of opening a bakery back home in his native Mexico.[46]

Antonio Perez survives with disabling injuries that are likely to be permanent. As mentioned previously, he and his wife and child were left without his income, and with mounting medical bills for the ongoing surgeries and care that he needed, and will need into the future.

And the fine instituted by OSHA? $79,200.

In an unusual move to be applauded, the State of Florida, Department of Financial Services, Division of Workers' Compensation fined Mac's Construction & Concrete, Inc. $2.4 million dollars in civil penalties for failure to provide workers' compensation insurance.[47]

As "Confined Space" pointed out, killing and permanently disabling workers cost a small OSHA fine. Not having workers' compensation insurance netted a $2.4 million fine, and a felony charge that carries a potential five years in prison.[48]

Because there was no workers' compensation insurance coverage, Mac's, et al. did not have tort immunity, and therefore could be sued.

Mr. Perez is reported to have won a $3.5 million settlement against Mac's Construction, et al.[49]

The Mac's Construction companies are now out of business. In January, 2006, the owner reportedly died at home of an accidental overdose.[50]

A Draconian 2003 change in the Florida workers' compensation law, pushed through by corporate and insurance interests, disallowed lawsuits unless the plaintiff could show employer intent to harm, or that employer had been warned previously that "a danger was virtually certain to cause injury or death to the employee," and that the employee was unaware of the danger. The latter defies everything that workers' compensation had originally been designed to prevent. It assumes that the employee has the power to force the employer to create a safe working place. Only among union workers in a strong union would there be likely to be any realistic possibility of an employee successfully confronting the employer on a safety/health issue. Florida had legislatively turned worker protection on its head.[51]

*   *   *   *

Author's Note: It took four requests over the course of one year, beginning six months after the Hobe Sound disaster, to receive OSHA Investigation No. 307300293—or part of it. Some parts were withheld; including the Forensic Engineering Investigation Report, witness interview statements, Tranquility floor plan guide, building permit, construction contract, jobsite meeting minutes, daily field reports, inter and intra agency correspondence, other. In March, 2006, an appeal was filed with the Solicitor of Labor's office requesting specific withheld documents, which were posted on April 27, 2007, or almost three years after the building collapse.

References:

1. *Death on the Job: The Toll of Neglect*, 13th Edition. AFL-CIO, Washington, D.C., April, 2004, p.32

2. United States Department of Labor, Bureau of Labor Statistics, Incidence Rates of Nonfatal Occupational injuries and illnesses by industry division, 1973–2002, cited in *Death on the Job: The Toll of Neglect*, 13th Edition, AFL-CIO, Washington, D.C., April, 2004, p.28.

3. Rotator cuff injury. Overview. Mayo Foundation for Medical Education and Research, www.mayoclinic.com, 8/25/04.

4. Definition of Rotator cuff, MedicineNet.com, www.medterms.com.

5. Supreme Court of the State of Florida. First Interim Report of the Fourteenth Statewide Grand Jury. Report on Workers' Compensation Fraud, Case # 90,703, filed 1/22/98, p.9.

6. All the above information cited from the Grand Jury Report is from the Supreme Court of the State of Florida. First Interim Report of the Fourteenth Statewide Grand Jury. Report on Workers' Compensation, Case # 90,703, filed 1/22/98.

7. ibid.

8. The names of the two deceased have been changed to protect their privacy and that of their families.

9. OSHA Inspection #307300293, Safety Narrative, p.2.

10. Martin County Sheriff's Office, Case No. 04–10322, Narrative Supplement #1; and OSHA Inspection No. 307300293, Safety Narrative, p. 2; and State of Florida Certificates of Death, #04 096193, and 04 096192 respectively.

11. Medical Examiner Reports, Case #s 04–19–484 and 04–19–485, Medical Examiner department, District 19, St. Lucie, Martin, Indian River and Okeechobee Counties, Florida, dated August 2, 2004.

12. The name of the injured worker has been changed to protect his privacy and that of his family.

13. Martin County Sheriff's Office, Case No. 04–10322, Narrative Supplement #1.

14. Fooksman, Leon; Tranum, Sam; Fernandez, Mireidy. Condo Collapse death toll at 2. South Florida Sun-Sentinel, July 24, 2004, p.l.

15. Florida Department of Financial Services, Division of Workers' Compensation, Bureau of Compliance, Initial Investigative Report, Compliance ID # 001394010, Investigative History, p.2.

16. Fernandez, Mireidy. No one pays as medical bills mount. South Florida Sun-Sentinel, 12/26/04, p.23A.

17. Case # 04–12575, 17th Judicial Circuit, Broward County, Florida.

18. OSHA Inspection No. 307300293, Safety Narrative, p.4.

19. ibid.

20. Margasak, Gabriel. Concrete linked to collapse. South Florida Sun-Sentinel, August 4, 2004, p.lA.

21. OSHA Inspection No. 307300293, Investigation Findings and Willful Justification, p.6.

22. ibid.

23. www. outinord-americas.com/tech__tunnel form.htm, 2006.

24. OSHA Inspection No. 307300293, Investigation Findings and Willful Justification, pp.6, 7; and Florida Building Code 1906.2.2.1.

25. OSHA Inspection No. 307300293, Investigation Findings and Willful Justification, p.7.

26. ibid.

27. ibid.

28. ANSI standard for Vertical Shoring, A10.9, Section 6.1.1, cited in OSHA Inspection No. 307300293, Safety Narrative, p.7.

29. OSHA Inspection No. 307300293, Investigation Findings and Willful Justification, p.7.

30. ibid.

31. ibid.

32. OSHA Inspection No. 307300293, Citation 01, Item 001, p.3.

33. OSHA Inspection No. 307300293, Citation 01, Item 002, p.l.

34. OSHA Inspection No. 307300293, Citation 01, Item 002, p.8.

35. OSHA Inspection No. 307300293, Citation 03, Item 001, p.20.

36. OSHA Inspection No. 306293507, Field Notes.

37. OSHA Inspection No. 306180134, Citation 01, item 001, 8/25/03.

38. OSHA Inspection No. 307300293, Inspection History, p.4.

39. OSHA Inspection No. 307300293, Citation 02, Item 001.

40. OSHA Inspection No. 307300293, Citation 01, Item 001, p.4.

41. OSHA Inspection No. 307300293, Citation 02, Item 001, p.13 and p.15 respectively.

42. ibid, p.15, 16.

43. Andreassi, George. Hiring private inspectors by developers questioned. South Florida Sun-Sentinel, July 24, 2004, p. 8A.

44. OSHA Inspection No. 307300293, Citation 02, Item 001, pp.16,17.

45. Fernandez, Mireidy. Falling concrete crushes hopes for a better future in Mexico. South Florida Sun-Sentinel, December 26, 2004,p. 23A.

46. ibid.

47. State of Florida, Department of Financial Services, Division of Workers' Compensation, Initial Investigative Report, Mac's Construction & Concrete, Inc., Control No. 04 227 D2, and Mac's Custom Construction, Inc., Control No. 04 225 D2.

48. Confined Space: Kill Workers = Small Fine; No Workers Comp = Go To Jail. November 24, 2004, cited at spewingforth.blogspot.com/2004/11/kill-workers -small-fine-no-workers.html.

49. _____ v. Macs Construction: $3.5 Million-Settlement, cited at www.hag-gardparks.com/notable__cases.asp.

50. Othon, Nancy. One condo collapse lawsuit still not settled. South Florida Sun-Sentinel, April 4, 2006, p. 3B.

51. Walker, Jessica. Construction Up, Injuries Up, But Workers' Comp Payouts Down. Daily Business Review, 10/20/05.

CHAPTER 7

▼

NURSES: EDUCATED, WRECKED AND CANNED[1]

Nursing professionals care deeply for your health and mine. As these stories show, that caring comes at a considerable price to their bodies and their lives.

Over 558,000 healthcare workers in the United States reported a work-related injury in 2001.[2] Nurses and other healthcare workers in hospitals, skilled nursing facilities, home care, and other personal care facilities have the highest incidence of work-related injuries and illnesses from lifting and other over-exertion activities of any industry.[3]

Nurses are desperately needed, vital to patient health and survival, yet are being lost to injuries that have been demonstrated to be completely avoidable; if hospitals and healthcare facilities provided the proper equipment and specialized lift teams.[4]

What follows are four personal stories representing two nursing professions, three different levels of care, and four widely geographically divergent states. All have the same unfortunate result—severe injury and financial loss, coupled with a wrecked career, permanent disability, and impaired quality of life.

*   *   *   *

ANNE HUDSON, RN, BSN

"How I'm Celebrating Nurses Week, May 6–12, 2003.

"I'm mourning the loss of my nursing career to a back injury from lifting patients. I'm not burned-out, fleeing the bedside. I'm an experienced nurse who wants to work as a nurse, but my employer will not take me back because I can no longer lift and haul patients. I'm grieving that ANA is not campaigning to protect nurses from preventable disabling injuries with Zero Lift for Healthcare legislation, and is not assisting injured nurses to keep their jobs, even during the shortage."

Anne Hudson wrote that statement in response to an email query from the American Nurses Association (ANA) asking "what you and/or your facility are doing to celebrate National Nurses Week."[5]

Hudson notes that shortly after, on June 21, 2003, ANA released their "Position Statement on Elimination of Manual Patient Handling to Prevent Work-related Musculoskeletal Disorders."[6] On September 17, 2003, ANA announced their "Handle with Care" initiative for prevention of musculoskeletal injuries in nurses. In early 2005, ANA sent their "Handle with Care" brochure, which includes Anne Hudson's story of work-related musculoskeletal injuries, to every hospital in America.[7]

Anne Hudson had completed her Associate of Science degree in Nursing at the community college in her small Oregon town, and entered the field of nursing when her sons were in their teens.

That was in 1990, and she went to work as a Registered Nurse on hospital medical/surgical, telemetry, and intermediate care units. Later, Anne went on to university, graduating with her Bachelor of Science degree in Nursing.

Hudson worked at the same acute care hospital for ten years, until her injury. She was truly committed to providing the best care and helping her patients to achieve the best outcomes. She loved hospital nursing and misses it now.

On the units that Anne Hudson worked, the patients typically require heavy care and have limited mobility. Many were unable to move themselves side to side in bed, or sit up, or transfer to a chair or bedside commode, without being physically pulled, lifted, or carried. If patients are both very heavy and too weak to assist, or are confused and therefore resist, the strain is increased, Anne explained.

There was only one mechanical lift at the hospital at that time. "You never knew which floor or unit the lift was on; and in order to use it, you had to call the nursing supervisor to request a sling for the lift and wait for the sling to arrive from central supply."

At the last of May, 2000, Anne Hudson was walking through her kitchen at home when she felt the pain—a sudden, horrible, low back pain so intense that she couldn't move. She initially thought she would be better and she would be able to go back to work that Saturday. On Friday, she called her physician's office. It was closed. When Saturday came, she was in too much pain; she had to call in, because she could not go to work. On Monday, June 3, 2000, Anne went to the hospital to fill out the workers' compensation form and to see her primary care physician, who initially diagnosed lumbar radiculopathy.

Along with awful low back pain, Anne had severe pelvic muscle spasms and intense leg pain from the spinal injury. She now had a cumulative trauma injury to lumbar discs. "My life as a nurse, as I had known it, was over. Other than a brief unsuccessful attempt a few months later, I have been unable to return to floor nursing."

All of the patient lifting that Anne did was contributing to the cumulative trauma injury. Anne calculated how much she was lifting during a shift, i.e., the weight of the patients, and how much she would lift to walkers, to chair, etc. during a shift. She calculated she was doing about 3000 pounds of lifting, pushing, and pulling in an average eight-hour shift. She thought this sounded excessive until a few months later, when she found an article from Australian nurses who had computed lifting 1.8 tons (3,600 pounds) per shift. Anne's estimate was actually low.

Since that initial diagnosis, Anne has been diagnosed with degenerative disc disease, lumbar strain, and bulging or herniated discs. She has been seen by a multiplicity of practitioners: neurologists, orthopedic surgeons, neurosurgeons, physical therapists, and a chiropractor.

Anne's primary care physician didn't send her for an MRI for a while. She "was not collapsing at that time or incontinent," so he was treating conservatively, she says.

Anne went to "Back School," defined as training in body mechanics, nutrition, drinking water to hydrate one's discs, etc. She tried heat and cold, exercises, lumbar supports and cushions, analgesic rubs, over the counter analgesics, and prescription anti-inflammatories, painkillers, and a muscle relaxant.

Because Anne's hospital was self-insured, and had not enrolled employees in a workers' compensation managed care organization, Anne was able to choose her own treating physicians and to get second opinions. When one is permitted to see only a workers' compensation-designated physician, one is likely to have a much harder time receiving timely and adequate treatment.

When Anne Hudson filled out the workers' compensation papers, the employee health nurse at the hospital worked out a plan for temporary modified duty in the nurse education department. Anne also did light duty in other areas. This involved some sitting and some moving around, but Anne was able to do it with the help of over-the-counter analgesics, a lumbar cushion in her chair and changing body position frequently. Anne could perform this light duty. She just couldn't lift. "So the hospital put me on the pathway", she says. "Ninety days of light duty, and if you can't return to the same heavy lifting, you are put on the pathway out the door." Over a span of two years, Anne was permitted only about 6 months of light duty and this led up to when she had surgery. Anne says she could have been working light duty the whole time, and wanted to, but there was no policy to do so.

After two months, workers' compensation denied her claim, stating it wasn't a work injury. Anne was injured about three months before she got an attorney. It was a blessed thing, Anne says, for the attorney was receptive to scientific research she had located on cumulative trauma spinal injury from heavy lifting and nurse back injury.

When her workers' compensation claim was denied, the hospital dismissed Anne from light duty and put her on Family Medical Leave. She went home, using her accumulated leave and her short-term disability insurance through the hospital.

She had her first MRI and was referred to a chiropractor. The first partial pain relief Anne had was 4 months after the initial pain, when she received spinal manipulation and ultrasound treatment from the chiropractor. Anne's chiropractor became pivotal in her care. He was first to recommend a muscle relaxer. He also insisted that she had disc pain, not just lumbar strain and recommended an MD to do a discogram. Anne explains that the discogram is a high-tech test used to confirm and examine internal disc injury.[8] "Discogram is done under fluoroscopy with injections of radio-opaque dye into spinal discs. It is used to establish whether a disc is the source of back pain, and, if so, which disc(s). People can have degenerative disc disease (DDD) that shows on MRI, but it may not be the source of pain. Discogram is the only test that examines internal disc disruption and can identify which discs are pain generators," explains Anne.

Anne has also had two series of physical therapy sessions after the injury, with heat, massage, ultrasound, and exercise; which she has found to be very helpful.

One physician told Anne that he had seen this many times over the years, that it is worse when it happens at work then at home, and that it's worse with job dissatisfaction. "If I were you, I would go find another job and put it out of your mind," he told her. And Anne began to think, "Well maybe I'm confused. I thought I loved my job. Maybe I'm not injured at all, but am having all of this pain just because I think I'm injured." She was trying to be the good patient. At the next visit, he told her there's nothing wrong with her back that could account for the pain in her legs.

Another physician talked about studies of identical twins in their 70s and 80s that showed no difference in their spines. The implication being that, therefore heavy lifting doesn't damage backs. This has been thoroughly refuted in many rigorous studies, notes Anne.

After a month or so, she saw her leave was being used up and was concerned for her job. She requested a prescription for a back brace and a release to return to work. By now it was October 2000, and Anne returned to intermediate care, wore the back brace with metal stays, and was taking analgesics. However, she had also been reading about back braces, and how they make you weaker because it's all external support.

Anne only lasted 3 weeks. Moving a 425-pound patient finished her off. Anne explained the back brace did prevent bending at the waist, and twisting. But when she lifted, the pain was severe because the brace did nothing to relieve the compressive force to the spine that occurs with lifting. Anne had the patient for three or four days, the patient was 5'2" and weighted 425 pounds. Anne is 5'3" and weighs 115 pounds. The patient was in hospital with a gastric bypass and was doing well post-operatively. When the patient assignment was made, Anne was reassured that she would have help moving the patient. But when Anne found the patient slumped down in bed, miserable, and needing to be pulled up, no help was available. Anne went to the head of the bed, and used the sheet to pull the patient up. And over the next days, Anne had to lift the patient's abdomen to wash under the folds. When the patient's bed was replaced with a recliner, Anne then had to bend down to provide care. Anne had to leave, her back was in too much pain.

Anne filed a second workers' compensation claim when she went off with increased back pain after caring for the 425-pound patient, and again while this claim was pending, she was assigned light duty. This time she was wearing scrubs, working in the pre-op holding area. It was wonderful to be working with patients again.

Thirty days later, Anne's claim was denied by workers' compensation, and the hospital sent her home.

Through a union negotiation, she was allowed another ninety days of light duty. Anne has urged that in the next contract, the nurses negotiate a "zero lift" policy requiring use of mechanical patient-lift equipment and a "permanent light duty" policy to keep back-injured nurses working. Anne stated that when she was injured in 2000, the attorney for the state nurses association told her that nurse back injury is "not a labor issue," that it is between the employee and the employer.

Anne knows nurses who have returned to work post-injury as unit secretaries at about half their previous RN wage. Anne says that if an RN can return to the same nursing position, with the same heavy lifting, that's fine. If not, it appears that most hospitals will move her/him out the door.

Workers' compensation closed her case prematurely, when Anne was preparing for surgery. She began the eligibility process for vocational rehabilitation, which determined that she was eligible. In the state of Oregon, Anne explains, one doesn't get to the decision regarding vocational rehabilitation until the workers' compensation claim is closed. She adds that they look at the community in which one lives and whether any other work is available at 80% of wages at the time of injury. Such work was not available, so she was found eligible for retraining.

Her attorney had to file papers in order to get her workers' compensation case re-opened for the surgery she needed. He succeeded, and her claim was reopened. Anne proceeded toward surgery, and did not pursue vocational rehabilitation. She did, however, obtain a copy of her vocational rehabilitation evaluation, which included a very long list of nursing "transferable skills."

Yet she and many other back-injured nurses like her, in hospitals across the country, states Anne, are not permitted to return, solely because they can no longer lift.

But what about the desperate need for nurses that we keep hearing about? The dumping of injured nurses doesn't appear to make sense.

Anne knows of one hospital that is hiring retired nurses, as a way of helping to solve the nursing shortage. They opened the positions to past employees first, for part time, mentoring, staff crunches, other areas that sound a lot like modified light duty. Anne emailed and asked that the project please consider injured nurses. "Many times, we can do all but the heavy lifting, and these are nurses who did not choose to leave nursing." During the nurse shortage, some nurses have chosen to leave the profession, but Anne points out that back-injured nurses are being forced out, as though their only valuable nursing skill is lifting. "Nurses are being injured, discarded and prevented from returning. Most of the injuries are entirely preventable, by use of safe patient-lift equipment and friction-reducing devices. Then after suffering unnecessary disabling injury, the doors are often barred to keep injured nurses from returning," states Anne. "Damaged goods don't sell," she quotes a back-injured nurse friend of hers.

How to prevent patient handling injuries, save financial resources for hospitals, and to examine the impact of preventable disabilities to healthcare workers is the subject of the book Anne Hudson co-edited with William Charney, entitled "Back Injury Among Healthcare Workers: Causes, Solutions, and Impacts."[9]

Since William Charney pioneered nurse back injury prevention in the early 1990s, Anne says, many studies have shown that the way to prevent patient handling injuries is through a zero-lifting policy, requiring use of patient-lift equipment by either specially trained teams, or nursing staff. Anne states that it is time to remove hazardous manual patient lifting from the backs of nurses and to implement modern technology designed for the task. The scientific evidence for nurse injury prevention is overwhelming. Hospitals are very slow in innovation.

"It's a very simple concept", says Anne, "If you remove the lifting activities that cause the back, neck, shoulder, and other injuries; you reduce the injuries. Through the use of technology, the stress is greatly reduced or eliminated, whether the lift equipment is used by nurses, or by the trained lift team. Also friction-reducing products, such as slide sheets, reduce the force of horizontal patient moves such as lateral transfers, pulling up in bed, and repositioning.

"A properly run lift team uses mechanical lift equipment and friction-reducing devices to perform lifts safely. We need a new term, such as 'no lift team' to indicate the difference: It is not reliant on the person, but rather on modern technology and training for safe patient handling. Studies have shown that when patient lifting is handed off to a trained lift team using lift equipment, nurses don't get injured, and neither does the lift team. One study (Meittunen, et al, 1999) reported 60,000 transfers without an injury."[10]

Anne talked about the high percentage of nurses (38%), who suffer back injuries. She cited research by Bernice Owen: A survey of 502 nurses showed that 38% of the nurses had had back pain or injury, severe enough to require time off work. Also alarming is that only 33% of injured nurses in the sample reported it as a work related injury.[11] The available data indicate that the incidence of work-related injury is greatly underreported.

It was the work of Bernice Owen, with the 38% injury rate, that first revealed to Anne Hudson the extent of the problem with nurse back injury.[12] It appeared to Anne that body mechanics training was ineffective in preventing nurse injury with patient lifting.

From nursing school, Anne remembers an instructor who told the class to take care of your back because your job depends on it; that nurses have a high rate of back injury, and are at high risk of cumulative trauma injury from lifting patients, but that workers' compensation only pays for single-point-in-time injury. "Your job depends on your back," she remembers. Anne wondered if the instructor meant risk of muscle strain, which could be expected to heal in a matter of days or weeks. The nursing class was not taught how and why cumulative trauma injury occurs to the spine itself from heavy lifting. Since her back injury, Anne has researched this information for herself.

"In England," Anne explained, "injured nurses are assisted to continue working. It was reported that an English nurse confined to a wheel chair from a motor vehicle accident, now works in an Accident and Emergency (which is equivalent to our emergency department). The Royal College of Nursing works with injured nurses to assist them, even supporting their cases in court. And the government backs the Royal College of Nursing in patient handling injury cases. Manual Handling Operations Regulations 1992,(MHOR) sets weight limits, so lifting any adult patient is precluded by the weight limitations. MHOR applies to all industries. In England, the nursing 'no lift' policy works in conjunction with MHOR."[13]

Australia, Anne continues, had national manual handling regulations for ten years, but nurses still had the highest injury rate in the female workforce, because the principles of safe handling of loads weren't being applied to patient handling. The Australian State of Victoria was successful in reducing nurse back injury only after nursing collaborated with government and healthcare on the Victorian Nurses Back Injury Prevention Project (VNBIPP). Costs for the $7.7 million project were recovered in one year by injury prevention, and Victoria is now saving $13 million per year in reduced nurse injuries.[14]

The American nursing community is fully aware of the physical damaging and discarding of nurses. Anne Hudson's theory is that, with nursing 95% female,[15] the reason this is allowed to continue is partly because it is a female problem in a male dominant society.

Anne notes that many US hospitals still promote 'correct' patient lifting. "This fosters the false notion that if a nurse gets injured lifting, it is the nurse's fault. Many hospitals do not tell the staff that lifting adult patients is never safe; and if they become disabled from the lifting, they are likely to be terminated. They do not adequately warn about cumulative trauma injuries." Anne explains that, "spinal discs lack pain receptors in the center of the disc, where damage from hazardous lifting typically begins. So a person can be sustaining damage over time but will not feel pain until the damage extends to the outer ring of the disc, where pain receptors are located. Also, nerves grow into fissures in the disc, the pH of the disc changes; chemicals leach out into the surrounding area and irritate nerves. Then the person experiences severe pain.

"The damage can be suffered over a long period of time, and pain may or may not happen at the time of the activity that has caused the damage. Many nurses, whether they have back pain or not, would show degenerative disc disease on MRI, the result of overuse injuries and not from spontaneous disease process. In my case, as a floor nurse, many outrageous patient lifts and moves contributed to my spinal injury."

On May 28 2002, Anne had surgery: a two-level, anterior-posterior, lumbar interbody fusion, with anterior placement of cadaver bone grafts, and posterior fixation with titanium hardware. Anne recently had "bending films" which showed that L5/S1 is fusing but that L4/L5 has not fused, which "may account for my continued pain," she says. "Nerves can grow into spaces that aren't supposed to be there. Right now, it's wait and see if L4/L5 fuses. That means more x-rays, with the cumulative risk of x-rays."

The hospital denied both of Anne's workers' compensation claims and appealed Workers' Compensation Court and Board decisions in her favor. Anne had to appeal the hospital's denials and go to court twice to prove that her injury was work-related.

Ironically, the evening of Anne's first court appearance fighting with the hospital was also the hospital's annual employee awards event. Anne Hudson received her Ten Year pin and accepted, with two other nurses, on behalf of the nurses' bargaining unit, the hospital's first ever "Team Spirit Award" for assisting with fund raising for the medical needs of a co-worker's young child.

Once Anne's leave had been exhausted, she no longer had benefits. She then purchased health insurance through COBRA (the Consolidated Omnibus Budget Reconciliation Act amendment to ERISA that allows the continuation of group health coverage that would otherwise be terminated), for which she was paying $938 per month at the time of this interview. Under COBRA, she can only purchase this insurance for a maximum of eighteen months.

Anne has applied for positions at the hospital, but as late as three years after her injury, and with permanent lifting restrictions, and her workers' compensation claim still open, they wouldn't hire her.

The worst part for Anne emotionally was that she was prevented from continuing her hospital career, when she demonstrated that she was well able to work apart from heavy lifting. "I thought they would keep me working because I was a dependable, faithful employee, a good nurse, and enjoyed good relationships with co-workers. Was I wrong!

"Seven months after my spinal injury, I went to work part time at a temporary, non-benefited job as a public health nurse consultant—great people, a pleasant and challenging job, very nice setting. Still, there are severe losses, which will never be regained. I'm making $10 less per hour and paying $938 per month for health insurance.

"After six months at the temporary job, I started a regular, full-time, benefited job as a public health nurse with the county public health department. It's very rewarding and stimulating work. I know how lucky I am, as a back-injured nurse, to be working as a nurse today. Yet, there are severe losses. When I should be at the height of my earning power, the gap is approaching $15 per hour between my current wages and what I would be making at my former hospital job."

About a year after she was injured, Anne learned that the hospital had a Back Injury Prevention Task Force. She requested permission to speak at their meeting, where she told them about the patient lift equipment that was available, and about the tremendous cost savings with injury prevention. They told her, "We know all that." So what Anne learned was that "the hospital could have prevented my injury and the injuries of others, but didn't."

*   *   *   *

REBECCA RHOADS, RN, BSN, CLNC

Half a country away, Rebecca Rhoads also suffered a debilitating, career-ending injury. Rhoads worked in an acute care hospital, on critical care units and in radiology. Since her injury, Rhoads has turned to writing and painting as a way to express the pain and the emotions of the experience. She has written her own story. It is as follows:

"I have a drawer that is filled with files, quite overfilled really. I call them my truth files. They represent a sizable chunk of my life, especially the last three years. I suppose the day I decided to go to nursing school is more accurately the beginning of the tale, but that would be way back to 1985. All those years strain the drawer so that it rolls slowly, weightily. This is fitting, for I also move slowly with the weight of those years.

"The injury that eventually led to my radically-altered life happened in 1998. At that time, I was working in the radiology department of a large, private, non-profit hospital. In my position as a registered nurse, my duties were wonderfully varied, and I enjoyed a great deal of autonomy. My years of experience in critical care nursing gave me the skills to make patient care decisions rapidly, expertly, and accurately. I loved my work and was very good at it. My employer agreed, giving me excellent marks at each annual review.

"I had had ten years of reviews in 1998; the skills I used were both cerebral and physical. It was the physical nature of what I did that left me disabled, physical duties to which I gave very little thought. The only time I thought about damage from the constant lifting, repositioning, and hefting of heavy bodies was when I attended the mandatory Safety Day, back safety being part of the curriculum. The back safety video all employees were required to watch presented information about the damage that could accumulate in the back over a period of years from lifting activities, inferring that using good techniques would prevent this damage. Conversely, if we were injured, it inferred we were to blame. I took the admonitions to heart and religiously used proper body mechanics. But patients don't come with handles. They move, shift weight, resist, and come in sizes too big, and much too big.

"Those physical activities were leaving a legacy in my body. The event that unleashed the senescent monster in my back happened in October 1998. One seemingly simple patient transfer unmasked the accumulated damage and left me bereft of my health and my profession. The diagnosis, degenerative disc disease, did not describe the severe damage discovered during surgery. What the diagnosis could not describe was the pain; the incredible, monstrous, burning pain, unrelenting, stupefying, bewildering pain in my low back. The diagnosis did not tell of the throbbing and aching pain down both legs; nor did it tell of the strange weakness and numb sensations that waxed and waned.

"Nurses are taught to care for others. We tough it out when we are sick or hurting, ever reluctant to see a physician. We keep working, even when our bodies signal that something is amiss. We hide pain, especially back pain: a back injury is the kiss of death for a nurse's career. She or he will deny the injury until the pain is so severe that simple walking is nearly impossible. Yet, nearly 40% of all nurses will suffer a career ending/changing back injury. The tragedy is these injuries are preventable, if the proper equipment is provided. A back injury is the most frightening thing that can happen to a nurse. So, we cope with a back injury in the most unreasonable, illogical ways. We think if we pretend it didn't happen, maybe it will go away. The psychiatric nursing books call it denial.

"I willed the injury to go away. It didn't. I went through a round of restricted work and physical therapy in 1999. The physician, a specialist in occupational health, didn't spend any of the hospital's money on an MRI to find out what was wrong, not even a simple lumbar spine x-ray. I was informed that perhaps I should look for another area in which to work; but was then released back to my same job, no restrictions, with no assistance to find a less physical position. No one explained my rights under the Workers' Compensation laws. Heck, I didn't even know I had a caseworker, who was supposed to help me with this. I was concerned that if I admitted I needed another position because of my injury, the hospital would seek ways to pressure me to quit. I'd seen it happen to other nurses. My fears were well founded.

"So, I plugged on for the next two years in the same job. I worked in pain. On days off, I rested until the pain abated, so I could go back to work again. Always the fear of losing my job was there if I but admitted I wasn't okay. Every time I lifted a patient, pushed a stretcher, or wore the leaded aprons, my back was weakening. Like a lumberjack hacking at a tree with an ax, the first blows only scar the wood. Each successive blow eats at the integrity of the mighty trunk, chipping away the years of growth. Eventually, the tree falls. I should have realized I was in trouble when I had to give up gardening and cleaning my house because of the pain. Those indicators should have motivated me to seek help, to admit to the pain. But the fear of losing my profession, my identity as a nurse, was the greater motivator. Oh how I wish I had listened to my body! How I wish I had saved myself and sacrificed my career! Yet, the damage was already there; insidious, snarling, waiting to manifest in unfettered fury.

"My colleagues? Some wanted me to leave the department. By January 2001, I was again seeing the doctor at the hospital's department of occupational health and medicine, whom I had seen two years earlier. (I use the term 'doctor' loosely—he was a hospital hireling, a 'shop doc.' In fact, he distorted and fabricated portions of my medical record—but that's another story.) He restricted me from wearing the heavy lead aprons. This was a problem for other nurses in the department, because it meant I could no longer perform on-call duties. Some became angry at me, at my injury,

"Fortunately, I had other colleagues who were supportive. They were willing to accommodate my problem. They understood it could happen to them and though troubled by that prospect, also understood how they'd want to be treated under similar conditions. I silently thanked God for their kindness, a commodity I found very little of in the months thereafter.

"Between January and the end of March, 2001, I had multiple interventions to attack the back pain. These included physical therapy, epidural injections, steroids, and narcotics. I had an MRI, and spinal x-rays. The pain worsened. I knew I was headed for the knife, knew I needed surgery. It terrified me. Subconsciously, I sensed my career in traditional nursing was over; yet outwardly, illogically, I hoped for a miraculous recovery.

"At the end of March, the pain made walking, sitting, or standing for any length of time unbearable. On a dark Saturday evening on March 24, 2001, my husband ferried me to an urgent care clinic. He stopped by the entrance to let me out. I hobbled, slowly, in excruciating pain, toward the entrance; managed to brace myself on the structures, and limped through the door. A clerk at the front desk saw me and immediately put me in a wheelchair. Sitting in a wheelchair is normally embarrassing for a healthcare provider; I had never been more grateful for a wheelchair.

"By this time, I could not work at all. The physician whom I saw, an emergency medicine specialist, recommended another MRI. I had had the first one at the end of January when it was apparent something was very wrong. I knew I needed another MRI. Anyone with common sense and knowledge of medical conditions would have reasonably concluded I needed another MRI. The hospital's occupational health doctor I saw two days later refused to order it. Had I been more astute, less trusting, I would have realized this refusal of service portended the coming odyssey of abandonment by my employer. An anesthesiologist, whom I saw for an epidural injection, ordered the MRI. By then, I had no left ankle reflex. He was disgusted it hadn't been ordered. It was nice to have an advocate.

"On April 19, 2001, I saw an orthopedic surgeon. As I had predicted, he recommended spinal fusion, the kind where huge titanium screws are drilled into the bony structures surrounding a fragile spinal cord and screaming nerves. This is just an everyday procedure for an orthopedic surgeon; but scary as hell for the patient, especially if she happens to be a nurse. There was a problem: The hospital's department of occupational health wouldn't approve the surgery and insisted I see a different orthopedic surgeon for an independent medical evaluation (IME) to opine if I really needed the surgery. I couldn't walk. I'd been confined to a couch for weeks because of the pain. My surgeon was noted to be extremely conservative as well as talented, and they wanted to know if I really needed surgery? I appealed, begged for approval, so that I would have a better chance for successful surgery. The longer a body is down before surgery, the more difficult it is to recover. It was to no avail. I spoke with an attorney. She said "for goodness sakes, see if your husband's insurance carrier will approve the surgery or just get it done—your health is at stake!" I had the surgery on May 22, 2001, paid by my husband's insurer. I was determined to get well again.

"Meanwhile, back at the hospital, another agenda was fomenting. Two weeks before surgery, I had a very unsettling conversation with my manager. Earlier in the year, I'd been assured that I was a 'valuable employee,' and that the department would find work for me to do, and would keep me no matter what. "You are a good employee—we don't want to lose you." This attitude radically changed when I was told my surgery wouldn't be approved, until I had the IME. Now, the good manager informed me he could post my job after I'd been absent for 12 weeks. I was devastated. I'd had the unsettling premonition that my career was over, but still I had hoped for the best. The message became very clear while I was hospitalized for the surgery: my husband's employer sent me a beautiful bouquet of flowers; my own employer sent nothing. One week after surgery, I was given the message that my workers' compensation claim was being disputed. The compensation abruptly stopped. So did my trust in my employer.

"The drawer full of files started about that time. I began to keep records of everything, copies of letters and policies, even notes for my attorney of every conversation and action taken by me and also by my new enemy, the hospital. This was the very place that was supposed to be for healing, the place where I had devoted my energy, my talents, my passion to helping others. Yet, when I needed the same compassionate help, it was withheld. During that summer while I healed from the surgery, I contacted my manager regularly. I explained my likely restrictions and expected date of return to restricted duty. Yet, when I was finally released to light duty, my job was posted. I received no prior notification. It was done with utter contempt and disregard for the devotion I'd shown as a nurse to the hospital. In September, I had a meeting with the director of my department. I was handed a letter that stipulated the conditions of my employment, rules I must follow. I wanted to tell him where to shove his letter and leave the hospital behind. But I showed up to work as the doctor had allowed, obviously shocking the director, who had expected his dirty work had been done, and I was out of his hair forever.

"Though I still had significant pain, I performed deskwork, knowing that MY JOB was posted, that other nurses were applying for it and that I had to leave when one of them accepted my position. That was stipulated in the 'Letter.' But when someone eventually did accept the position, the hospital reneged on its end of the bargain! Suddenly, they wanted me to stay until she started. So when she started, I understood that I was to leave. But the goal post had been moved again!

They wanted me to stay until her orientation was completed! I drew the line when I was asked to train her. Obviously, the 'Letter' only established that I was under the hospital's beck and call, and they could change the conditions of my employment at will. I made call after call, and sent my resume to several departments seeking alternative work. The doors, if there had been any, were locked shut. No one wanted a nurse with a bum back even for paper work, especially if she'd made a workers' comp claim. It was retaliation, punitive and ugly.

"In January, 2002, the nurse who had been hired into my position had completed her orientation. The manager told me he didn't have any job for me. He said somberly that he'd check into how my employment was to be handled. But brightening, he said he'd look for something in another department. He would help me find something. It was so magnanimous! It was so considerate! It was all rotten carrots dangled from a very long stick!

"I was sent by the occupational health and medicine(OHM) manager to an interview for a job in the business office. I'd been given a copy of the e-mail sent to the OHM manager from a manager in the business office. The e-mail described the position, asking the manager if she knew of a nurse whom they might interview. When the same person who had originated the e-mail asked me how I found out about the position, I told her I'd been sent by so-and-so at OHM, the very *one* to whom she had addressed the e-mail. She looked at me in puzzlement, asked who this person was, and was quickly 'reminded' by the director of the business office that it was so and so. She replied, 'oh, that so-and-so person." The interviewer's focus on the source of the referral, over the position itself, confirmed what Rebecca Rhoads already knew. "The job was a red herring."

"During the interview, I was told how much a nurse was needed in this role, especially a nurse with my particular background. I was cautioned that the job would be posted as a temporary job, and the permanency of the position would depend upon how much money in lost charges were recouped during the trial period. The position was posted the week after my surgeon placed me back on medical leave because of worsening pain and the possibility that I was facing more surgery. I found out a year later that the poor nurse who was suckered into this position was dismissed after the three-month trial. The position was determined to be a clerical position. I guess they really didn't need a nurse after all. This same tactic, this same position had been used other times to lure injured nurses into what seemed a good job, only to have it abruptly restructured as a clerical position. Clerks make far less per hour than registered nurses.

"I interviewed for a different job, one that I knew I could do. I'd been told that someone else had already been offered the job, even before it had been posted..... The interview was utterly insulting, hostile. The department director walked in over 10 minutes late. Questions were repeated, though I'd adroitly addressed them the first time asked. The interviewers most eagerly assured me my back disability was not a problem, even though I was pointedly questioned about how long I could sit. The assurance was just a little too quick. They'd obviously been coached what to say and what not to say. The interviewers were nurses with whom I'd worked, people who knew me. They acted as if I'd just walked in from Antarctica. The atmosphere was precisely as cold as the Antarctic. In the end, 'another candidate' was selected, my former co-worker. I knew she did not have her baccalaureate degree, otherwise our experience was nearly the same. She was without a disability though. All RN job descriptions in the hospital include wording that encourages attainment of a bachelor's degree, even emphasize the importance of education. All, except this one. The deck was stacked.

"Now that my replacement was finished with orientation, it was time for me to leave the Department. My last day was January 19, 2001. I was told by payroll that in order to go from active to inactive employment status, a certain paper had to be filled out and signed by my manager. When I went to co-sign the paper the next week, it said "termination due to physical disability." I went into near shock. Termination? This is what I get after 13 years of dedicated service? I asked the manager twice. Is this a termination? He shrugged and said he didn't know what else to do. I walked out utterly befuddled, then had the presence of mind to turn around, go back and request a copy of the paper. I read and re-read the paper. I'd never been terminated from any job. It was humiliating.

"I had retained an attorney who specialized in Workers' Compensation to handle the dispute. I consulted him after mulling over what to do about being terminated and was given another attorney's phone number. From there, I was directed to seek help from the State of Michigan Department of Civil Rights. Eventually, a kind gentleman, a field agent, initiated an inquiry on my behalf. A month after the termination, I was magically reinstated! It was all a big mistake! The termination was "rescinded"—but I still had no job. The letter informing me of this came with apologies, and with papers to be filled out to place me on medical leave. The hospital was calling all the shots and all I could do was comply, for fear that any misstep would negate my workers' comp claim when we finally went before the magistrate.

"Two years later, my case went before the magistrate—and I won! The dirty laundry was flown from the clothesline. The doctor who fabricated portions of my medical notes; the wrongful terminations,(there were 2); my fruitless, disheartening attempts to find favored work; the day I mentally, emotionally crashed and burned after my job was posted behind my back; all was displayed. It was ugly and the judge had the wisdom to understand what I'd been through, that it was unconscionable and that my injury was legitimately work-related. Even the defense's doctor testified during a deposition that my injury was work-related! The workers' compensation dispute was shameful, orchestrated only because the hospital gambled that I'd either be back to work or that I'd simply walk away. They expected I'd walk away if they made it hard enough on me. I summoned all the grit God had given me and persevered. Had I walked, I would have ruined my claim for compensation.

"So, that's how the drawer began to fill with files, files of letters, documents, medical notes, exchanges of information, and my exacting documentation. The papers are emotionless pieces of me. They cannot describe the pain of loss, the pain of suffering through yet another back surgery because the first fusion had failed. I blame the failure squarely on the useless, cruel stress the hospital put me through. A body cannot heal when a heart is broken and beating in pieces, when the spirit is wounded and without hope.

"I've learned much in the last two and one-half years. I have come to understand much about workers' compensation, especially how easy it is for unethical employers to violate the civil rights of employees. The system provides NO equal protection, although that is our constitutional right. Injured employees become citizens of a third-world system that not only fails to protect the injured; it allows the employer to bankrupt the employee physically, emotionally, and financially. Workers' compensation should not be used as a ruse for increasing an employer's revenue. The hospital where I worked is self-insured. They get money back from the fund, if they don't spend it on the injured employee. In 2001, their workers' comp dispute rate was nearly 60% as compared to an average 17% of the state's self-insured hospitals with comparable numbers of claims. It is an investment scam.

"Nurses specifically need protection from injury, as well as from unethical employers. Healthcare workers need ergonomic standards to prevent injury. England and Australia have zero-lift policies. When nearly 40% of professional nurses suffer a preventable, career-ending/changing back injury, many of those permanently disabled; I wonder why this legacy continues in an era of shortage of experienced hospital nurses. Ironically, it is often the best nurses who are injured, because they are the very ones who are the most conscientious about patient care.

"My files exist to protect me from a ruthless and unconscionable business. It has discarded all pretense of magnanimity for disgraceful, hateful treatment by the employer of the employees who make it the hospital that it is. It is as if an injured employee becomes a leprous parasite; a calculated business decision is made to cull the leper. That was all I had become, a business decision. If I were anything more than a business decision, if one shred of humanity remained in the hospital that directed my misery, those who make those decisions could not possibly sleep at night. I sleep on a pillow of a clear conscience and for now, until I endure and prevail through a long appeals process the hospital's attorney has promised, that is my reward."

ADDENDUM to Rebecca Rhoads

"Sadly, the files continue to grow. In October, 2004, I was diagnosed with arachnoiditis (ARC). ARC is considered a rare condition involving the spinal meninges, specifically the arachnoid layer. The meninges are tissues that surround and protect the brain and spinal cord. The arachnoid layer produces fluid that cushions and nourishes the brain and spinal cord. ARC results mostly from things that doctors do, including spinal surgery. Because it is largely iatrogenic (physician caused), little research has been done to find treatments for ARC. In my case, blood that collected around the spinal cord, during surgery in 2001, caused the condition. Blood cells cannot cross the meninges. However when blood is broken down in the body, numerous chemicals are released that do cross into the arachnoid. These chemicals are the cause of the initial inflammation in the arachnoic layer.

"Normally, nerves move and are flexible; so whenever a leg is moved or a person bends, the nerves slide and give. In ARC, the nerve roots are scarred, hardened, and adhered to each other, and to the meninges. The nerves are not only inflexible but also pull on other nerves and the dura (outermost meningeal layer). This same scarring can prevent the normal circulation of the cerebral spinal fluid. If the fluid becomes backed up, it causes the dura to enlarge in the area where bone is removed during certain spinal surgeries. As one can imagine, this all causes considerable pain. ARC sufferers have a shortened life span, and a significantly higher suicide rate than the general population. It is considered a debilitating, incurable, albeit rare, disease. Treatments are directed at pain management primarily, but also assisting sufferers to maintain some physical activity to prevent further deterioration. (The interested reader may see www.arachnoiditis.com for further information.)

"I first realized I had arachnoiditis when I conducted an internet search and stumbled upon a well-researched paper published on-line, describing the disease and its causes.[16] I was stunned when I realized the description precisely fit my symptoms and surgical history. MRI studies done in September 2004 confirmed what I already knew: I had ARC, a permanent, incurable condition. The disease is debilitating and incredibly painful. Few physicians know about ARC. As a nurse, I had never heard of it. The disease started after the first fusion surgery, as nerve root clumping is present on the MRI studies of February 2002, but not in the MRIs done before then. Thus, it is apparent that the disease was present as early as February 2002, but went undiagnosed.

"I now see a world-renowned anesthesiologist and pain management specialist, who has become an expert in diagnosing and treating sufferers of ARC. As his office is several states away, he works with the local physiatrist whom I have been seeing for pain management since 2002. Both physicians have been a godsend to me. Many others with ARC are less fortunate, and are given the royal brush-off by less skilled and empathetic care providers. Because I am in the chronic phase of the disease, there is nothing to do but to maximize my medication regimen and remain as active as possible within my limitations. I now have to use a wheelchair whenever I need to walk about in a park, or large garden, a museum, or even a large city.

"Was nursing for 13 years worth the price of developing an incurable, debilitating disease? I was ignorant of what could happen to me physically while I was contributing to my patients' well being, even saving their lives. Nursing was worth the investment of my intellect, and the patients were worth my time, my effort, my very best efforts—but not my body. My employer was worth none of these."

by Rebecca Rhoads, RN, BSN, CLNC[17]

As late as September, 2005, four years later, the self-insured hospital still had not paid back wages and medical costs from the date of the re-injury in 2001 through late 2003. In 2003, the magistrate's decision ordered the hospital to pay from that point onward, pending appeal.

State of Michigan has a Second Injury Fund, which becomes liable after 52 weeks of disability for a certified "vocationally disabled employee" who is injured on the job.[18] The purpose of the fund is to encourage employers to hire workers who have suffered a previous disabling injury by reducing the employer's liability for re-injury to 52 weeks. However, it also requires timely notice by the insurance carrier to the fund in the event of re-injury, which Rebecca tells me, did not happen in her case.

Now that Michigan Supreme Court has decided a precedent setting case, Rebecca Rhoads explains, her case can now be heard.[19]

*   *   *   *

It is not only hospitals that are physically dangerous places to work, but skilled nursing facilities and home care, as well:

CAROL ANN STOUGHTON, LPN

Carol Ann Stoughton had been a nurse for 23 years. Hers was a mid-career change. She had gone back to school in her late 30s, doing very well and winning acclaim in her studies. Carol was good at nursing and loved what she did.

As a private duty Licensed Practical Nurse(LPN) at the time of her injury in November, 1999, Stoughton was working with a child who needed everything done for her. Neither the family nor the employer provided a lift or any assistive devices. Lifting the 60-pound child from the shower bath into the bed tore the rotator cuff and ligament in her right shoulder.

National Institute for Occupational Safety and Health(NIOSH) recommendations limit lifting to 51 pounds, i.e., that 51 pounds is the heaviest object one can safely lift.[20]

In the previous year, one other caretaker for the child had also been injured, but still no lift equipment had been provided.

Stoughton reports it was four months after the injury before her employer's workers' compensation insurer approved her for an MRI, and another 4 months before Carol was seen for evaluation by an orthopedist. Over that eight-month period, the pain and increasing atrophy in her shoulder would cause a second injury, while she still lifted the child, now relying on her other side to do the work, and injuring her left elbow. The MRI revealed "significant spurring at the acromioclavicular (AC) joint, and findings consistent with impingement secondary to the AC joint hypertrophy, and also suspicious for rotator cuff tear of the shoulder." The rotator cuff tear was confirmed on physical exam.[21]

The rotator cuff is made up of muscles and tendons that connect the upper arm bone with the shoulder blade. It also holds the ball of the upper arm bone in the shoulder socket. The shoulder is designed specifically for movement and has the "greatest range of motion of any joint in your body."[22]

Workers' compensation kept Carol Stoughton waiting 14 months for the surgery. Physical therapy was attempted, and Carol progressively got worse. When surgery was finally performed, it debilitated her even more and left her with pain, adhesions, numbness, tingling, and immobility. Carol says she later learned that this workers' compensation-approved physician's expertise was not shoulders. What she now knows is that the surgery will need to be performed again, properly this time.

Both before and after the surgery, the agency kept her employed. The how was the shocking part. Carol was assigned to a small office with a card table, and a shredder that could only handle one page at a time. Her job now became shredding file box upon file box of paper. At first, she had a card table to lean on, then they took that away; and she had a chair, file boxes and a shredder. It was here that she received her third injury. Lifting up files from the floor damaged her wrist and injured her hand.

More and more employers are putting injured or ill workers on restricted work, instead of sending them home to recover. Between 1992 and 2001, the number of reported cases with days away from work declined by 34%, while the number of cases with days of restricted work increased by 64%.[23]

Over the next few years, the muscle loss from atrophy, the pain, the destruction of her career and her livelihood took its toll on Carol Stoughton, both physically and emotionally.

Carol had to fight and finally obtain an attorney to get the wage compensation checks that would give her only a percentage of her previous earnings. Even then, the checks would be late, or they would stop. Everything was a fight with the insurer. "Workers are treated inhumanely," states Carol.

Carol had planned on retiring at age 68 years. Now her career has been cut short with consequent loss of earnings, as well as added costs.

Carol Stoughton feels the differences in her life. The limited range of motion and the degenerative changes in her right shoulder give her difficulty, even in such ordinary tasks as combing or washing her hair, putting on and taking off clothing. She is aware of the difference in her appearance caused by the atrophied muscles and out-of-alignment shoulder.

Carol applied for Social Security Disability five times before she received it. This is not an unusual circumstance, I was to learn, as I talked to many disabled workers around the country. At the time we spoke, Carol was awaiting her first check. Her injury, which totally disrupted her livelihood, was deemed to be a 12% injury.

Carol Stoughton settled her workers' compensation case for an amount that will never come close to the lost wages and lost social security, had she worked as she had planned. And the cost of redoing the operation on her shoulder will likely absorb most of the settlement, she says.

Yet Carol feels fortunate that her injury came toward the end of her wage earning years. As she says, she didn't lose her house. But she misses who she was, knows she can't be there again.

Carol had been a very physically fit woman before the injury, following a family tradition. She is the daughter of two gymnasts, both of whom are now in their 90s. Her father was an Olympiad several times.

The most devastating part, explains Carol, is that if proper medical treatment had been provided in a timely manner when the injury occurred, she would not have the scar tissue, adhesions, and atrophy that have caused so much pain, immobility, and deterioration.

At the time we met, Carol had a second visit scheduled with a vocational rehabilitation specialist in a few days, to see what she might be able to do to earn a living without using her body. This too, has been a series of delays and ineptitude, with Carol having to wait 8 months for the required case study to determine how, if at all, she can receive services. Apparently, Carol explains, the Florida State Vocational Rehabilitation Services periodically runs out of money and has to wait until the next fiscal year to approve certain kinds of services in support of re-training or re-employment. But Carol lights up as she starts to think of ways she might still be able to use her nursing background.

This bright, energetic woman, tough-minded and resilient, is simultaneously deeply saddened by what this injury has done to her life. After a continuous battle, Carol lives on a percentage of her former income. Once very physically fit, she sees a physical deterioration she can't stop.

If proper lift equipment and assistive devices had been in the home, Carol would not likely have been injured.

As Carol Ann Stoughton says, "If the employers and the insurance companies would treat us humanely, effectively and simply 'right,' along with keeping safety factors first and foremost, everyone would benefit."

ANASTASIA, LPN

At the other end of the country, in New York State, Anastasia had hoped to work until 72 years of age to make up for the early years when she was raising her four children with "nothing going into Social Security." Instead her career was cut short at age 61, when the last of four falls and injury to her knee occurred. Anastasia did not want her full name used.

Anastasia left a marriage to a very controlling man who wouldn't allow her to work. Her previous skills in payroll had been rendered obsolete by computers. While Anastasia went to work, she also went to school nights to become a Licensed Practical Nurse (LPN). That was 18 years ago. Anastasia would have preferred to get her RN, but neither time nor finances would allow.

Anastasia worked as an LPN on the Alzheimer's unit, in a skilled nursing facility (SNF). SNF patients rely on physical assistance many times throughout the day and night, into and out of the bath, on and off the toilet, moving into and out of bed, into and out of chairs, turning in bed, walking, if able, etc. Patients who can't assist, or are combative or agitated, all of which occurs routinely on an Alzheimer's unit, compound the muscular-skeletal stresses in patient handling. Anastasia was injured four times over as many years.

The first time Anastasia was injured was in April 1992, as she and several aides were lifting a patient into the special chair for the monthly weighing. They had no lift or other equipment. Midway through lifting, the male patient resisted; and Anastasia twisted her leg, injuring her right knee so badly it required surgery. Anastasia had asked the nursing home to provide an orderly to assist with moving patients, but without success.

The workers' compensation carrier delayed an MRI for months. Then Anastasia had to wait another month for approval for the surgery.

The second injury occurred in February 1993, as she and an aide were attempting to clean an Alzheimers patient who was allergic to milk, but would manage to get milk and drink it anyway, and then get diarrhea. The patient constantly paced and Anastasia and an aide were attempting to change the patient. The patient was fighting them and there were feces on the floor. Anastasia slipped on the soiled floor and landed on the same knee. Again, the injury required surgery. This time, it took three months for workers' compensation to approve the needed surgery.

In December 1995, Anastasia fell on the ice at the change of shift as she was coming into work, injuring the same knee. A few months later, she hit a stool in a storage room, and the knee locked on her. Surgery was again performed.

Since this final re-injury on the same knee, Anastasia has been unable to work as an LPN. She asked for re-training in another field, but workers' compensation refused, suggesting that she do home care or day care, even though both these nursing venues would present the same problems.

Throughout these years, Anastasia almost lost the cooperative apartment she had managed to save enough to buy. For a long time, workers' compensation gave her nothing and she had no funds. At one point she got an eviction notice from her co-op and was saved by a friend who loaned her the money.

Now she is just barely making it financially. The injuries cut short her career and her ability to save for retirement.

Anastasia now uses a cane and a brace, the latter if she is going to walk a block or more. It is heavy, she says, referring to the brace.

Workers' compensation fought and delayed until just recently. Even now, she says, they claim they overpaid her in 1992, and want to take money back. She has saved every stub, everything regarding the workers' compensation. And she adds, "They talk so down to you".

It took a union coming in, to get assistive devices into the skilled nursing facility, Anastasia explains.

*   *   *   *

It is a dreadful irony that healthcare facilities disregard the health and safety of their own workers. This arrogance and reckless disregard comes at a high price in dollars, worker lives, and patient outcomes. We know there is a better way.

From a practical perspective alone, it makes no sense to lose nurses, or other healthcare staff, due to disabling injuries, thus compounding an already critical nurse shortage.

Patient mortality has been shown to increase dramatically when the number of hospital patients per nurse was increased.[24]

"Mechanical lift equipment could prevent most nurse and patient injury with lifting. But nursing staff are being used as disposable human lift equipment, many being fired when 'broken' by unsafe lifting. This abuse must be stopped by legislation which prohibits forced hazardous manual patient lifting," states Anne Hudson.[25]

American Nurses Association, in concert with state nurses associations, has developed a nationwide state legislative agenda on safe patient movement and handling to promote legislation to protect nurses and patients from injury with manual lifting.

In October, 2005, Texas became the first state in the United States to pass legislation that mandates a policy of safe patient handling. Texas Senate Bill 1525, which the Texas Governor signed into law, requires the governing body of a hospital or nursing home to "adopt and ensure implementation of a policy to identify, assess, and develop strategies to control risk of injury to patients and nurses associated with lifting, transferring, repositioning, or movement of a patient."[26]

In March, 2006, Washington State's governor signed into law a bill that mandates the provision of lift equipment as part of a hospital's policy for safe patient handling.[27]

In January, 2007, United States Representative John Conyers, Jr., Michigan District 14, re-introduced Nurse and Patient Safety and Protection Act of 2007, HR 378, "To direct the Secretary of Labor to issue an occupational safety and health standard to reduce injuries to patients, direct-care registered nurses, and other health care providers by establishing a safe patient handling standard."[28]

One can hope that an effective national mandate to protect health care workers and patients will be realized soon.

References:

1.   The title for this chapter is credited to Anne Hudson.

2.   U.S. Department of Labor, Bureau of Labor Statistics,Bureau of Labor Statistics Data at www.bls.gov.

3.   Musculoskeletal Disorders(MSDs) and Workplace Factors: A Critical Review of Epidemiologic Evidence for Work-Related Musculoskeletal Disorders of the Neck, Upper Extremity, and Low Back, Chapter 1. U.S. Centers for Disease Control, National Institute of Occupational Safety and Health, www.cdc.gov/niosh/ergtxtl.html.

4.   Charney, W., Hudson, A., Ed. *Back Injury among Healthcare Workers: Causes, Solutions, and Impacts.* Lewis Publishers, CRC Press, 2004.

5.   email Anne Hudson to American Nurses Association (ANA), dated May 7, 2003, in response to ANA query, "What you and/or your facility are doing to celebrate National Nurses Week", May 6–12,2003.

6.   Position statement on elimination of Manual patient handling to prevent work-related musculoskeletal disorders. American Nurses Association. June 21, 2003. www.nursingworld.org/readroom/position/workplac/pathand.htm.

7.   Handle With Care. American Nurses Association, Silver Spring, MD, in cooperation with Patient Safety Center of Inquiry, Veterans Health Administration, Tampa, FL. The production and distribution of this brochure was

funded by a grant from Johnson & Johnson. 2004. www.nursingworld.org/handlewithcare/brochure.htm.

8. Note: Discography is recognized by many physician specialists as an invaluable procedure that provides data that can't be determined any other way. However, for some, it is controversial.

9. Charney, W., Hudson, A., Ed. *Back Injury among Healthcare Workers: Causes, Solutions, and Impacts.* Lewis Publishers, CRC Press, 2004.

10. Meittunen, E.J., Matzke, K., McCormack, H., Sobczak, S.C. The effect of focusing ergonomic risk factors on a patient transfer team to reduce incidents among nurses associated with patient care. Journal of Healthcare Safety, Compliance & Infection Control. 3 (7), August/September, 1999.

11. Owen, B.D. The magnitude of low-back problems in nursing. Western Journal of Nursing Research. 1989, 11, 2; 234–242.

12. Owen, B.D. Preventing injuries using an ergonomic approach. Association of Operating Room Nurses Journal, 2000, 72 (6), p. 1031–1036.

13. Manual Handling Operations Regulations 1992, derived from European Council Directive (90/269/EEC), www.hmso.gov.uk/acts.htm, cited in Hudson, A. UK Patient Handling Practice and the Manual Handling Operations Regulations 1992.

14. Victorian Nurses Back Injury Project Evaluation Report, 2002. Policy and Strategic Projects Division, Victorian Government Department of Human Services, Melbourne, Victoria, Australia. October, 2002. www.nursing.vic.gov.au.

15. The Registered Nurse Population. Findings from the National Sample Survey of Registered Nurses. Health Resources and Service Administration. Bureau of Health Professions. Division of Nursing. US Department of Health and Human Services. March, 2000. p.39.

16. Smith, Sarah. The Arachnoiditis Syndrome. BackCare. National Organisation for Healthy Backs, East London & Essex Branch, Chelmsford, Essex, UK, cited on www.btinternet.com/~thinkback/arach.htm.

17. Rebecca Rhoads, RN, BSN, CLNC wrote her own account of what happened to her under workers' compensation.

18. State of Michigan, MCL 418.901a, 418.921.

19. Bailey v. Oakwood Hospital & Medical Center. State of Michigan Supreme Court case Number 125110, filed June 29, 2005.

20. Applications Manual for the Revised NIOSH Lifting Equation, Appendix 1, Publication 94–110, CDC, NIOSH, Washington, DC, January, 1994.

21. MRI Report, 3/15/00, cited in Medical Evaluation for Carol Stoughton, dated July 16, 2001.

22. Rotator cuff injury. Overview. Mayo Foundation for Medical Education and Research, www.mayoclinic.com, 8/25/04.

23. *Death on the Job: The Toll of Neglect*, 13th Edition. AFL-CIO, Washington, D.C. April 2004, page 5.

24. Aiken, L.H.; Clark, S.P.;Sloane, D.M.;Sochalski, J.; Silber, J.H. Hospital Nurse Staffing and Patient Mortality, Nurse Burnout and Job Dissatisfaction. Journal of the American Medical Association, JAMA, Vol. 288, No. 16, October 23–30, 2002.

25. Statement by Anne Hudson in an email to American Journal of Nursing, November 8, 2005.

26. Texas Senate Bill No. 1525, to amend State of Texas, Section 1, Subtitle B, Title 4, Health and Safety Code, Chapter 256.002 (a), cited in Hudson, Mary Anne. Texas Passes First Law for Safe Patient Handling in America: Landmark Legislation Protects Healthcare Workers and Patients from Injury Related to Manual Patient Lifting. Journal of Long-term Effects of Medical Implants, 15 (5) 559–566. 2005.

27. www.leg.wa.gov/ub/billinfo/2005- 06/pdf/Bills/House%20passed%20Legislature /1672-S.PL.pdf. and anne@wingusa.org.

28. HR378; Nurse and Patient Safety and Protection Act of 2007, @http://Thomas.loc.gov/

CHAPTER 8

▼

MEDICAL TREATMENT
UNDER WORKERS'
COMPENSATION

Rodrigo Aguilera injured his back and right leg when he was struck by an electric forklift at work on April 21,1999. At the hospital emergency room, blood in the urine was also identified.

Inservices, Inc., the workers' compensation benefits company hired by the employer, referred Aguilera to a workers' compensation clinic "where he was treated and discharged to return to work with restrictions." Shortly after, Aguilera began to complain of bladder and kidney pain. Through his attorney, he "requested examination and treatment by a board certified urologist. Inservices denied the request, claiming the injury was not work-related." In June, 1999, when "Aguilera notified Inservices that he was passing feces through his urine and was in need of immediate urological care," he was notified that his workers' compensation benefits were being terminated, despite the report of two physicians that he should not return to work. Inservices also denied the prescribed medication. Five days later, on June 30, 1999, "the emergency request for care by a urologist was denied, ostensibly because Inservice deemed it "not 'medically necessary.'"

Several weeks later, Aguilera's treating physician "again advised Inservices that the need for urological care was 'urgent' and that his condition had deteriorated." Aguilera's physician ordered various urinary tests and scheduled appointments, only to have some of those appointments cancelled by Inservices. A urethogram that was performed, however, revealed that Aguilera had a "fistula or hole in his bladder."

On August 19, 1999, Aguilera's attorney notified the adjuster that Aguilera "needed a general surgeon to perform emergency surgery on the fistula." The new case manager assigned by Inservices "refused to authorize emergency surgery and insisted on a second opinion." Despite having been instructed not to deal directly with Mr. Aguilera, she showed up at the office of the urologist performing the independent medical examination at the time of Mr. Aguilera's appointment, and then asked Mr. Aguilera to lie to his attorney and say "that she was not at the doctor's office."

Inservices insisted on Aguilera undergoing tests "that were painful ... and contraindicated by his ... medical condition." Inservices then "used Plaintiff's (Aguilera's) refusal to submit to these painful tests as a further excuse to refuse Plaintiff's now critical, surgical treatment."

"After urinating feces and blood for over ten months," and after being "forced to be seen by six doctors in addition to his initial treating physician," who also "concluded that his physical injuries were in fact related to the injury and that his condition as a result required urgent surgical treatment," Aguilera's surgery was finally authorized in late March 2000. Aguilera had been left in pain, in a health threatening, life threatening condition for ten months, battling all the while to overturn the dangerous, egregious treatment he was receiving at the hands of the workers' compensation insurer.

In this landmark Florida Supreme Court case, the court found that the carrier's outrageous conduct during the claims process "would preclude application of statutory immunity."[1]

The abuses in medical treatment under workers' compensation are legion and varied. Themes that run throughout the experience of injured workers are the delays; the denial of treatment; the inadequate, inappropriate and cursory medical treatment; inhumane treatment; and the shunting of costs that should be paid by workers' compensation to others, including the injured workers and their families.

Herbert, et al.(1999) studied 135 patients diagnosed with work-related carpal tunnel syndrome between 1991–1994, who had filed for workers' compensation and were being treated at a New York-based occupational medicine clinic in an academic institution. Fully 79% of these patients' workers' compensation claims were initially challenged or received no response. Once adjudicated, however, over 96% were accepted as work-related. The mean time for a claim to be adjudicated was 429 days, with a maximum of 1617 days. The mean time for treatment authorization was 226 days, extending up to 1296 days, or 3.6 years. The mean time for authorization for surgery, following a physician's request, was 318 days, with a maximum of 595 days. "Claims filed by non-whites, low-wage workers, and union members were significantly more likely than others to be challenged."[2] The likely impact of such lengthy delays, in addition to pain and decreased mobility, is to reduce the effectiveness of treatment; thereby increasing the likelihood of poorer treatment outcomes and permanent disability.

The average time to resolve a disputed claim in California, based on an analysis of 26,400 cases, is 60 months, or five years, up 23% over the past decade.[3]

In many cases, the very same employer/insurer who is battling the injured worker every step of the way is also selecting the physician on whom the worker is relying for proper diagnosis and appropriate treatment. Some corporate employers provide their own on-site medical treatment. The best medical interests of the injured worker are thus compromised by the physician's allegiance to the corporate employer or the workers' comp insurer. In one instance, a company classified its on-site treatment as first aid, in order to circumvent OSHA reporting requirements.[4]

In a plastics molding plant, a sworn affadavit submitted to OSHA from a former employee claims that she was under orders not to record any injuries that didn't require hospitalization or a minimum of three physician appointments or didn't cost more than $200 to treat. The employee reportedly was fired a month after raising concerns about the OSHA logs with her boss. The company claimed she was fired for not keeping proper logs, but the logs after her dismissal appear to continue to show injuries that should have been recorded and weren't. The company denies any intentional under-reporting.[5]

At this same plant, Landis Plastics, four workers reportedly had one or more fingers amputated within fourteen months. In March 1996, Landis is said to have admitted that four of its presses had not been adequately guarded, and promised to add electrical switches to shut off the machines when the guard wasn't down. But more amputations are reported to have occurred after. OSHA appears to have fined Landis for "deliberately failing to report 63 workers' injuries ... along with failing to safeguard machines...." The plant's injury severity rate was reported to be "more than seven times the national average" for plastics molding plants.[6]

Another common theme in reporting of work injuries is to "blame the vic-tim." In one Midwestern tire manufacturer, a worker slipped and fell in an icy company parking lot. The accident report identified the cause of the accident as "Worker's eyes not on path." It neglected to point out that the parking lot and sidewalk had not been cleared of ice and snow for several weeks, even though workers were required to use it periodically.[7]

Study after study in states across the country found significant under-report-ing of worker injuries and occupational diseases requiring medical treatment.[8] The US Bureau of Labor Statistics (BLS) data has been much criticized for its substantive under-reporting, not only of worker injuries and illnesses, but also of worker fatalities.[9]

In a study of occupational injuries in Athens County, Ohio for the five years from 1982 to 1986, Fingar, et al.(1992) compared hospital emergency depart-ment data for the two hospitals in the area, and lost-time (1 or more days) work claims reported to the Bureau of Workers' Compensation (BWC) for the same community. Of 6173 injuries, the workers' compensation system only identified 25.5% of the cases. Hospital emergency department records identified 81.2% of the work-related injuries. Only 410 cases showed up in both systems.[10]

In Connecticut, Morse et al. (1998) interviewed a random sample of working adults, identifying 292 residents with work-related upper extremity musculoskel-etal disorders. Of the 292, only 31 (10.6%) had filed a workers' compensation claim. Of the 31 cases, only 23 cases (7.9% of the 292) were accepted by the insurer. Sixty percent of cases were the primary wage earner. Another 13% had been the primary wage earner prior to the injury. Twenty-one (7.2%) reported job loss due to their condition. The authors note that physical and occupational therapy were more likely to be paid out of workers' compensation than other medical costs and procedures. The study concludes that almost 90% of likely work-related musculoskeletal disorders are not reported to workers' compensa-tion.[11]

Frumkin, et al.(1995) interviewed 107 persons with occupational injuries, 31.9% of the patients seen for occupational injuries at two Philadelphia emer-gency departments serving 17 poor census tracts during the base period, 1987–88. One-third of the workers were employed in health care, one quarter in the service sector, the remaining in construction and other. The interviewees were those who could be reached by telephone two to three years later (the time of the interviews), and agreed to participate (95%). Almost 40% of those interviewed reported continuing health problems resulting from the occupational injury. Of the 107, only 34 (31.7%) had applied for workers' compensation. Almost one-third, 11 claims, were reported contested by the employer/insurer. Ultimately, only 27 received workers' compensation, representing 25% of the sample.[12]

In two large automobile plants, Fine et al.(1984) found that medical record-based incidence rates for acute trauma were 4–5 times greater and for cumulative trauma disorders of the upper extremities were 68–93 times greater than incidences coded from the OSHA 200 log (the required work injury/fatality reporting instrument).[13]

These studies also demonstrate how a very low percentage of worker injury costs are being covered by workers' compensation. If workers' compensation is paying only a small percentage of the cost of worker injuries, one must ask how these costs are being paid?

In the same Connecticut study referenced earlier, Morse et al. found that 70.9% of medical visits and procedures were paid by general health insurance, with another 8.1% being paid out of the patient's pocket.[14]

Reville and Schoeni (2003/2004) found that 36.3% of adults ages 51–61 years who are disabled, and 28.9% of all persons with a disability receiving social security disability, are disabled due to work. Yet, only 5.3% ever received workers' compensation. The cost in Medicare and Social Security Disability for workplace injuries/illnesses is estimated at over $33 billion annually in 2001 dollars.[15] The medical component of workers' compensation in 1996 was estimated at $24 billion annually.[16]

A former California Insurance Commissioner, John Garamendi, is said to have proposed a unified health care system incorporating the medical component of workers' compensation into traditional group health, in order to control health care costs. It would avoid duplication, thus increasing administrative efficiency and generating savings that could be applied toward universal coverage in the state. Workers' compensation insurers fought hard against it. They were worried about losing control over the medical portion of the workers' compensation premium, and its accompanying big bucks. The bill died in the state legislature.[17]

The serious underreporting of work-related injuries and diseases in BLS and OSHA data has other consequences, particularly in the ability to identify and treat occupational diseases, especially those with a long latency period, such as toxic exposures.

Another factor in the failure to correctly diagnose occupational diseases is the scant attention paid to training of physicians to identify occupational illness. For the 1991/1992 academic year, the amount of time devoted to teaching occupational health/illness issues, in those US medical schools that include it in the curriculum at all, was an average of 6 actual hours.[18]

Milton, et al.(1998) reviewed 67 medical charts for patients in a Massachusetts HMO with adult, or late, onset asthma. Fourteen cases were identified as "attributable to occupational exposure." None were treated under workers' compensation. Of the 14 work-related cases, only two had been asked about their work by the treating physician, and in neither case were work-related symptoms noted by the physician. Workers' compensation paid for none of the 14 cases, nor were any of the cases reported to the state's program for recording occupational risks.[19]

In California during the 1980s, although almost two-thirds of the physician-reported occupational illnesses were eye and skin problems, "almost no occupational cancers were reported."[20]

In cancer cases, Dr. Samuel Epstein, Professor Emeritus of Environmental and Occupational Medicine at University of Illinois School of Public Health, Chicago, sees a pattern of "blame the victim," i.e., by blaming smoking, aging and/or genetics for cancer. He cites NIOSH surveys that as many as 15 million workers "are exposed to cancer-causing chemicals and radiation on the job, and that as much as 40% of cancers may be occupational."[21]

Qualified occupational medicine physicians are in limited supply. At the same time, this is the expertise that "should be required of either treating or consulting physicians, when the illness or injury is unusual or rare outside the workplace...."[22]

But even when a proper diagnosis is made, corporate and other pressure may be brought to bear on the physician.

This appears to have been the experience of one board certified radiologist at an upstate New York hospital who had relocated after years of experience at a large New York City hospital. He began to see the distinctive signs of asbestosis in the lung x-rays that he was reviewing. Week after week, he was seeing classic signs of asbestosis. As he explained, asbestosis is so unique on x-ray that it can be nothing else.

Some hospital physicians wanted to attribute it to smoking. Others diagnosed it incorrectly as emphysema, which as this radiologist noted, shows up on an x-ray as an over-expanded lung, the opposite of what you would see in asbestosis. These were miners; and what this radiologist found was that no one was willing to stand up for them, and some were dying.

He reportedly sent 50 cases to the state health department. The state had his diagnoses confirmed by a specialist at an occupational medicine institute.

What followed was an examination of x-rays of persons forty years and older in six hospitals in two counties. The team found 500 chest abnormalities, 71% were "indicative of asbestos exposure."

The radiologist's contract with the hospital was reportedly cancelled just before renewal.[23]

Other medical specialists have found similar results for their concern for their patients. In Libby, Montana, radiologists who are said to have first raised the alarm that vermiculite miners at a W.R. Grace and Co. mine had asbestosis, reportedly left town soon after.[24] Ten thousand people a year die from asbestos-related illnesses:[25]

Injured workers may risk adverse consequences when they attempt to identify a medical problem as work-related.

Pransky et al.(1999) interviewed 110 workers, managers, health and safety personnel at three industrial facilities in Massachusetts. What the authors found was that less than 5% of workers had reported a work-related injury or illness, although fully 85% had experienced work-related symptoms, with 50% experiencing continuing work-related problems due to their injury or illness. Only thirty percent (30%) had reported any lost time due to their work-related injury or illness. Workers gave "several reasons for not reporting their injuries, including fear of reprisal, a belief that pain was an ordinary consequence of work activity or aging, lack of management responsiveness after prior reports, and a desire not to lose their usual job." Some worried that they would jeopardize pay raises, or lose overtime, or promotions, or status. There were administrative barriers to reporting, as well, such as a desire to achieve a management goal of no reported injuries, and corporate incentives for lack of reporting, such as bonuses for reduced or no recorded injuries.[26]

Biddle, et al.(1998) analyzed almost 30,000 reports to Michigan's Department of Public Health of known or suspected work-related illnesses or injuries, filed in 1992–1994, against claims data from Michigan's Bureau of Workers' Disability Compensation (BWDC), to determine which had filed claims for workers' compensation wage replacement benefits. Claims for medical benefits only were not included in the BWDC database. Their findings: "between 9.6% and 45.6% of workers reported to have occupational injuries or illnesses filed for workers' compensation wage-replacement benefits." Workers with carpal tunnel syndrome were most likely to file for benefits, while workers suffering chemical or heavy metal toxicity were least likely to file. The latter illnesses are ones with a long latency period between the time of exposure and the development and progression of the disease. The authors suggest that many employees may choose to "'tough it out' or 'work through' the problem ..." rather than to file for workers' compensation benefits.[27]

In a cross-sectional study of 1598 workers in Michigan, primarily unionized autoworkers, diagnosed with work-related musculoskeletal disorders, Rosenman, et al(2000) found that only 25% of workers filed workers' compensation claims for their injuries. Those with the most severe injuries were most likely to file. Of those who filed, 13.6% received no benefits. Of the workers who did not file, 59% felt their injury was not serious enough, 28% expected to miss work but knew they would receive sick leave or short term disability from their employer, 36% believed their medical expenses were covered by other insurance, and 20% didn't think their injury was work-related. Those who were treated by family physicians or specialists were more likely to file for workers' compensation, than those who were seen by the company physician."[28]

Workers who file workers' compensation claims often find themselves in a Catch-22: Workers' compensation delays or denies the claim and thus does not pay for treatment, while medical insurance refuses to cover treatment because it is a work-related injury. In 2001, a Harvard study found that medical problems accounted for about half those filing for personal bankruptcy, representing 2 million persons, including dependents. Astonishingly, over 75% had medical insurance at the onset of their illness.[29]

Some employers have initiated programs that give cash or other prizes to workers who do not report injuries. Others have created disincentives, such as loss of overtime opportunity or an enforced day off without pay. One major food services company was reported to have developed a point system to penalize workers who file a workers' compensation claim, i.e., filing a workers' compensation claim would cost a worker one point; if the claim involved days away from work, the cost would be five points. Termination occurs at thirty points. All of these methods have one common thread: blame the victim. Instead of examining the hazards in the workplace and eliminating them, the focus is on blaming the victim and stopping the reporting. OSHA did a review of the literature and found that "providing rewards for fewer injuries chills employee reporting of injuries, rather than improving workplace safety; and reduces the reliability of the data on injury logs."[30]

Workers who did file had much to say about their experience with the workers' compensation brand of medicine:

- "The company doctor was a quack.... [H]e had people working under medication that says, 'Do not operate a vehicle.' And that's what we do. We drive for a living. 'Take this but you can go to work.'"

- "The clinic took x-rays of me. The doctor did not tell me, 'you're injured,' or 'you've sprained it'—nothing. He only touched me, he sat me in a chair, 'Wait here.' I laid down on a table. They put hot water compresses on me. Afterwards he told me, 'You need to go to therapy.' That was all the information that I had."

- "When it comes time to settle, you're broke. I had to take $1200 offered to give up future medical treatment. I took it."

- "The small amount they give you when the injury is severe just doesn't compensate, because the money runs out soon, and you will always be hurt."

- "Nobody can fix my back.... [H]owever bad your pain, your pain will continue. I would rather not get anything and just (be able to) do my work."

- "When I finally got to the point where I couldn't work anymore, the employer didn't want to acknowledge it. They wanted to fire me."

All the above statements are those of workers who participated in one of the eight focus groups of injured workers who had applied to the California Workers' Compensation Appeals Board.[31]

Another injured worker described his workers' compensation medical treatment experience as "demeaning" and being "treated like cattle."

The workers' compensation system puts the responsibility on the patient to prove that he or she is sick or injured, and that treatment is needed. At that point, treatment might be delivered, but in the cheapest way possible, whether that temporarily ameliorates the situation, or even if it creates a delay that reduces or eliminates the possibility of successful treatment with full recovery.

As Carol Stoughton sees it, "Workers' comp writes the book for the doctors to read on 'How to Treat the Claimant': discourage, dismiss, refuse and alienate...."

The Broward School District is one of the largest school districts in the nation, encompassing Broward County, Florida within the one district, with a staff of 39,000 teachers and other workers. The district had reportedly audited its workers' compensation system, described as managed by one company, which in turn, subcontracts medical duties, including selecting medical doctors to a California firm. Auditors chose five doctors at random from the subcontractor's list. One physician reportedly had four malpractice settlements since 1992.

Physical therapy referrals "were based not on a therapist's track record, location near the patient, or expertise. Instead, therapists were chosen alphabetically, based on which firm was next on an approved list," auditors are reported to have found.

The subcontractor is also reported to have "referred some patients to general practitioners rather than specialists," resulting in a worsening of some patient problems, "until they needed much more serious and expensive treatments, such as surgery."

One physician appears to have turned in claims primarily for treatment related to orthopedic issues, but his certification was in internal medicine and pediatrics.[32]

*   *   *   *

Even among the flagrant mistreatment that characterizes medical treatment under workers' compensation, the case of Dr. Ignacio Magana was outrageous. One might be tempted to see this as an extreme example, unlikely to be repeated. The point is that the workers' compensation system in most states has no "fail-safe" mechanism by which to protect an injured worker from being victimized by the Dr. Maganas of the workers' compensation world.

Dr. Ignacio Magana was a Harvard Medical School graduate, completed internship and residency at University of Miami, Jackson Memorial Hospital, and received specialty certification in neurological surgery.[33] He was reportedly Chief of Staff at Palm Beach Gardens Medical Center and on staff at the Neurosurgery Clinic, with offices in Palm Beach Gardens, Port St. Lucie and Vero Beach, Florida. His specialty was surgery of the spine and spinal cord. He had apparently also been a Clinical Associate Professor in the Biomedical Science Program at Florida Atlantic University.[34] Magana had a wife and three children,[35] volunteered on cub scout camping trips sponsored by a church cub scout troop.[36] In other words, Magana was a stellar and respected, upstanding citizen of his community—or so it appeared.

In December 2001, that all began to unravel. Within a week after Dr. Magana's arrest on sexual battery charges, at least ten women came forward, all alleging sexually inappropriate behavior in Dr. Magana's offices. One complaint went back to December 1996. When asked why they didn't report it when it happened, the answer was the same: "they didn't expect to be believed."[37] Only a handful of cases were within the time limit to be prosecuted.

On July 1, 2004, Dr. Ignacio Magana pleaded guilty to one charge of aggravated assault with intent to commit a felony. He was placed on probation for five years to be served concurrent with two other cases.[38]

The long delays in coming to the police, years in some cases, and the lack of physical evidence had hampered the prosecution.[39]

Magana's license to practice medicine was "relinquished" following disciplinary actions and he is unlikely ever to practice medicine in State of Florida again.[40] But how many women had to suffer at his hands before that happened?

*   *   *   *

The very fact that in medical treatment under workers' compensation one has extremely limited, if any, choice of medical providers; that change of physician is made difficult and lengthy; that the patient is not the customer; that the patient must rely on what the physician reports, accurate or not, for the treatment (s)he so desperately needs; and that insurers often put pressure on physicians to give them what they want to hear, through various means of persuasion and arm-twisting; all contribute to the potential for abuse.

In 22 states, the employer chooses the treating physician. In an additional 19 states, the worker's choice of physician is restricted. In only 9 states, does the worker have the right to choose his/her own physician.[41]

Sending patients to physicians at great distances (500 miles in one case in this book), not because they have specialized expertise, but because they are known to respond the way the insurer wants; delays in approval for treatment; denials for appropriate treatment; insufficient therapy visits to meet the patient's needs; lack of physician expertise in occupational exposures; and the fact that workers' compensation medicine is a separate entity, outside the normal medical pathways; all these work against effective quality medical care for injured workers.

In addition, the current system of separate reporting and under-reporting of work-related diseases is a severe detriment to epidemiological data gathering, aggregation and surveillance essential to assist in diagnosis, treatment, improved prognosis, and ultimately, prevention of occupational diseases. There is a need for a national comprehensive health data collection system for occupational injuries and illnesses.[42]

There is no need for a separate health care system for workers—except to fill the coffers of insurers.

References:

1. The preceding case information is from Aguilera v. Inservices, Inc., SC 03-368 Supreme Court of Florida, June 16, 2005, and Inservices, Inc. v. Aguilera, 837 So.2d 464 Florida Third District Court of Appeal, filed December 26, 2002,.

2. Herbert, T., Janeway, K., Schechter, C. Carpal tunnel syndrome and workers' compensation among an occupational clinic population in New York State. American Journal of Industrial Medicine. 1999; 35:335–342.

3. Fricker, M. Insult to injury. Debilitating delays. Press Democrat, (Santa Rosa, CA) 1997 www.pressdemo.com/workerscomp/day2/main2.html.

4. Azaroff, LS, Levenstein, C, Wegman, DH. Occupational Injury and Illness Surveillance: Conceptual Filters Explain Underreporting. American Journal of Public Health, 2002:92:9, p. 1421–1429.

5. O'Brien, J. In the past year, four workers have lost fingers while working at the Solvay plastics plant. The Post-Standard, Syracuse Newspapers, August 29, 1996, B1.

6. O'Brien, J. Feds want $720,700 fine for Landis: OSHA also may ask for a criminal probe into allegations the Solvay plastics company deceived the government. The Post-Standard, (Syracuse New York), January 15, 1997, p. A1; and Landis Plastics of Solvay, N.Y. Agrees to Pay OSHA $425,520;Make Multi-Facility Safety and Health Improvements. OSHA National News Release, 6/18/98,OSHA Archive, www.osha.gov.

7. Frederick, J., Lessin, N. Danger on the Job: The Political Economy of Worker Safety. Blame the Worker: The rise of behavioral-based safety programs. Multinational Monitor, Vol 21: 11, November 2000.

8. Azaroff, LS, Levenstein, C.,Wegman, DH. Occupational Injury and Illness Surveillance: Conceptual Filters Explain Underreporting. American Journal of Public Health, 2002:92:9, p. 1421–1429.

9. Leigh, JP., Markowitz, S., Fahs, M., Landrigan, P. *Costs of Occupational Injuries and Illnesses.* University of Michigan Press, 2000, p.216.

10. Fingar, AR, Hopkins, RS, Nelson, M. Work-related injuries in Athens County 1982–1986: a comparison of emergency department and workers' compensation data. Journal of Occupational Medicine. 1992; 34; 8:779–787.

11. Morse, TF, Dillon, C, Warren, N., Levenstein, C, Warren, A. The economic and social consequences of work-related musculoskeletal disorders: the Connecticut upper-extremity surveillance project (CUSP). International Journal of Occupational and Environmental Health. 1998; 4:209–216.

12. Frumkin, H., Williamson, M., Magid, D., Holmes, J.H., Grisso, J.A. Occupational Injuries in a Poor Inner-City Population. Journal of Occupational and Environmental Medicine, 1995; 37:12, 1374–1382.

13. Fine, L J, Silverstein, B A, Armstrong, T J, Anderson, C A. An alternative way of detecting cumulative trauma disorders of the upper extremities in the workplace. Proceedings of the 1984 International Conference of Occupational Ergonomics, 1984; 425–429. Cited by Pollack, ES., Keimig, DG. Counting Injuries and Illnesses in the Workplace: Proposals for a Better System. Washington, DC: National Academy Press. 1987: 136, cited in Azaroff, LS, Levenstein, C, Wegman, DH. Occupational Injury and Illness Surveillance: Conceptual Filters Explain Underreporting. American Journal of Public Health, 2002:92:9, p. 1421–1429.

14. Morse, TF, Dillon, C,Warren, N., Levenstein, C, Warren, A. The economic and social consequences of work-related musculoskeletal disorders: the Connecticut upper-extremity surveillance project (CUSP). International Journal of Occupational and Environmental Health. 1998; 4:209–216.

15. Reville, R.T., Schoeni, R.F. The fraction of disability caused at work. Social Security Bulletin, 65:4, 2003/2004, www.ssa.gov/policy/docs/ssb/v65n4/v65n4p31.html.

16. Himmelstein, J., Rest, K. Working on reform: how workers' compensation medical care is affected by health care reform. Public Health Reports, 1996; 111: 12–24.

17. ibid.

18. Burstein, JM, Levy, BS. The teaching of occupational health in US medical schools: little improvement in 9 years. American Journal of Public Health, 1994; 84:4: 846–849.

19. Milton, DK., Solomon GM., Rosiello, RA., Herrick, RE. Risk and incidence of asthma attributable to occupational exposure among HMO members. American Journal of Industrial Medicine 1998; 33:1–10.

20. Landrigan, PL., Markowitz, S. Current magnitude of occupational disease in the United States. Estimates from New York State. Ann. New York Academy of Sciences, 1989; 572:27–60, cited in Azaroff, LS, Levenstein, C, Wegman, DH. Occupational Injury and Illness Surveillance: Conceptual Filters Explain Underreporting. American Journal of Public Health, 2002:92:9, p. 1421–1429.

21. Dr. Samuel Epstein, Chairman, Cancer Prevention Coalition, cited in *Death on the Job: The toll of neglect*. 13th edition, AFL-CIO, Washington, D.C., April 2004, p. 15.

22. American College of Occupational and Environmental Medicine. ACOEM's Position Statement: Eight Best Ideas for Workers' Compensation Reform. Committee Report, www.acoem.org

23. Schneider, A. Special Reports. Pushing for asbestosis study cost doctor his job. Seattle Post-Intelligencer, June 22, 2000.

24. ibid.

25. Environmental Working Group, cited in Baker, P. Bush wants asbestos cases settled. The Washington Post, reprinted in South Florida Sun-Sentinel, January 8, 2005, p. 17A.

26. Pransky, G., Snyder, T., Dembe, A., Himmelstein, J. Under-reporting of work-related disorders in the workplace: a case study and review of the literature. Ergonomics, 1999, 42:1: 171–182.

27. Biddle, J., Roberts, K., Rosenman, K.D., Welch, E.M. What percentage of workers with work-related illnesses receive workers' compensation benefits? Journal of Occupational and Environmental Medicine. 1998; 40:4: 325–331.

28. Rosenman, K.D., Gardiner, J.C., Wang, J., Biddle, J., Hogan, A., Reilly, M. J., Roberts, K., Welch, E. Why most workers with occupational repetitive trauma do not file for workers' compensation. Journal of Occupational and Environmental Medicine. 2000; 42:1:25–34.

29. Himmelstein, D.U.; Warren, E.; Thorne, D.; Woolhandler, S. Market-Watch: Illness And Injury As Contributors to Bankruptcy. Health Affairs. Health Tracking Marketwatch. February 2, 2005, http://content.healthaffairs.org/cgi/content/full/hlthaff.w5.63/DC1, and Harvard study finds medical bills push many into bankruptcy. Harvard Law School, February 3, 2005, www.law.harvard.edu/news/2005/02/0 3_bankruptcy.php.

30. Frederick, J., Lessin, N. Danger on the job: The political economy of worker safety: Blame the worker: The rise of behavioral-based safety programs. Multinational Monitor. 21:11, November, 2000.

31. Sum, J., in consultation with Stock, L. Navigating the California Workers' Compensation System: The Injured Worker's Experience. An Evaluation of services to inform and assist injured workers in California. Prepared for the Commission on Health and Safety and Workers' Compensation. University of California at Berkeley, Labor Occupational Health Program. July 1996. Complete report available at www.dir.ca.gov/chswc/navigate/navigate.html.

32. Hirschman, B. Workers' comp audit cites waste. South Florida Sun-Sentinel, June 22, 2005, p.IB.

33. Practitioner Profile, State of Florida Department of Health, ME 47340, ww2.doh.state.f1.us/irm00profiling/profile.asp?.

34. Greenlee, W. Doctor accused of molesting patient. Port St. Lucie News, May 10, 2002, p. Al.

35. Motion to Mitigate, State of Florida v. Ignacio Magana, Case No. 02–2738-MMA, Circuit Court, 19th Judicial District, Martin County, FL.

36. Letter of support for Dr. Magana written to Judge Hershey included in court documents on Case No. 02–2738-MMA, Circuit Court, 19th Judicial District, Martin County, Florida.

37. Greenlee, W. Arrested surgeon has more accusers. Port St. Lucie News, May 15, 2002, p. Bl.

38. State of Florida v. Ignacio A. Magana, District Court, 15th Judicial Circuit, Case Numbers 02–006675CFA02, 02–006676CFA02, and 02–006677CFA02.

39. Neurosurgeon avoids jail time for sex charges. By Associated Press, Naples Daily News, July 3, 2004, www.naplesnews.com/npdn/florida/article/0,2O71, NPDN_14 910_3008868,00.html.

40. Practitioner Profile, Florida Department of Health, ME 47340 ww2.doh.state. fl.us/irm00profiling/profile.asp?.

41. Information from US Department of Labor, State Workers' Compensation Laws, 2001; and Workers' Compensation Research Institute, Managed Care and Medical Cost Containment in Workers' Compensation: A National Inventory, 1998–1999; cited in AFL-CIO Workers' Compensation Comparisons, 2001, www.aflcio.org/safety/compcomp.htm.

42. Azaroff, LS, Levenstein, C, Wegman, DH. Occupational Injury and Illness Surveillance: Conceptual Filters Explain Underreporting. American Journal of Public Health, 2002:92:9, p. 1421–1429.

CHAPTER 9

▼

MY OWN WORKERS'
COMPENSATION JOURNEY

On March 31, 1999, I was on my way back to my office when I slipped on a dirty floor at the hospital at which I worked. I landed on my hands, then onto my left hip. Almost immediately, I experienced pain in my hands, arms, shoulders, and numbness in my left arm. Thus began my personal chronicle with workers' compensation. My workers' compensation agreement does not permit me to name the hospital, or the health system of which it is a part, thus I will refer to both as Egregious General Hospital.

I couldn't get up for several minutes. Next, I became worried that someone else could come along and slip—now on my spilled coffee, combined with whatever else had been on the floor and upon which I had slipped. Ernst ___, from the hospital's maintenance department, rounded the corner, and seeing me on the floor, assured me he would stay there until someone from the hospital's environmental services department came to clean it up. I collected myself, got up and went to my office.

Most of my career has been as a health, hospitals, and/or mental health administrator. At the time of the injury, I was director of Egregious General and its' health system's pediatric developmental evaluation clinic. It was a step down from my previous position as regional VP for a psychiatric healthcare management company. That company had gone bankrupt. I had expected to move on quickly, but it was not to work out that way. The injury would turn out to be the end of my career.

After a short while in my office, the pain was now also in my neck, head, and left side. I went to the hospital's emergency department, which, upon seeing my hospital identification badge, immediately steered me out of emergency and into employee health. An RN at employee health gave me an anti-inflammatory medication and told me I'd probably feel worse tomorrow, and if I felt I needed physical therapy (PT) to come back.

The next day, I did feel worse, and stayed home. I noted that the medication was making me drowsy, therefore I could not use it and drive to work. The following day, I accepted the referral to PT.

That Monday I went to my PT appointment where I was pulled, pushed, and probed in the cervical spine (neck) area. My pain increased. I came away with my head reeling. What was clear to me was that the hospital was dangerously putting the treatment before the diagnosis; and there was no physician oversight of the treatment, a violation of hospital regulations and standards.

As I was to learn, employers and insurers are often willing to provide initial physical therapy; but will fight and deny diagnostic and other necessary treatment and/or surgical procedures.

I called employee health about an x-ray. I was told nothing would show up because this appeared to be a soft tissue problem. However, MRI was not offered, either.

I called my primary physician. When I told her there was no physician overseeing my treatment, she recommended I see a neurologist and gave me the name of her recommendation. I called his office, learned he didn't take workers' compensation cases, but his partners did. I called their office and confirmed: Yes, they took workers' compensation and could see me Monday, or earlier, if I needed.

But Egregious General would not approve my physician's recommendation. The fact that these neurologists accepted workers' compensation wasn't enough. "Too costly," I was told by the hospital's appointed gatekeeper for workers' compensation. "It could cost $600." Naively, I asked how much they would expect to pay. "Maybe $24" was the answer. They would give me a list of neurologists that I could use. After many days and two more phone calls, I finally received the list.

I called the five names from Egregious' list that my primary had recommended. One was retired, three would not take workers' compensation, although this was the workers' compensation list; and one had an earliest appointment one month away.

Employee Health steered me to the representative for the insurance claims adjuster (hereinafter called the insurer) that the self-insured Egregious General Hospital used. The insurer representative attempted to deny what I had just told her: No, I couldn't be right, this is a new list, etc. Then she asked why I was calling and what did I expect her to do. I explained I was in pain and an appointment one month away wouldn't do. I also told her my physician had recommended a neurologist who takes workers' compensation, who could see me by Monday. She wasn't hearing any of it.

By this time, we were eight days out from the accident. Employee health made arrangements for MRI and x-ray at the hospital that day. When I went to radiology, they told me they would page me when they were ready for me, so I wouldn't have to wait. Radiology paged me. Arriving a few minutes later, they told me it was a mistake; but could or would not give an explanation. I would have to call Central Scheduling, they said. Central Scheduling told me they could fit me in 12 days from now. I explained I was in pain. Go to radiology (where I had just been), they said. Perhaps they can get you in sooner. Sooner turned out to be 6 days away. Later they called back to say that, too, was a mistake.

I had been commuting 3 hours a day for 4 days in a row by this time. Driving was an agonizing experience, at the same time that it produced numbness in my left leg and arm. That afternoon, I got in my car and headed home; knowing I had a $1^{1/2}$ hour drive in pain and numbness, before I reached home.

I called my husband as I got close to home. When I arrived, he had to pull me out of my car, and assist me upstairs to bed.

Early the next morning, we went to the emergency department at the community hospital near our home. X-rays showed straightening in the cervical spine. Treatment by an orthopedist was recommended. The MRI, performed that afternoon, showed disc bulge at C4-C5, and C5-C6 (the 4th to 5th and 5th to 6th cervical spine discs) .

Upon completing the MRI, I called the insurer-approved orthopedist and was scheduled for 10 days away. In the meantime, I was on painkiller, and muscle relaxant, and on leave from Egregious General, until I started treatment. In reality, I worked from home, parts of all but 5 days. A medical transport service brought me to and from one all day meeting. Medical transport, however, would not have been useful most days; since I had both meetings in the community and responsibilities at the hospital, and the transport would only pick up and deliver to one place.

Those five days, in keeping with hospital policy, came out of my earned time off. If the hospital logged those 5 days as earned time off, did it also mean that the hospital excluded this injury from the hospital's required BLS Survey of Occupational Injuries and Illnesses, and the OSHA 300 Log of Work-Related Injuries and Illnesses? If so, this debilitating, and what I now know to be permanent, injury would not have even been recorded as a work injury requiring time off from work.

There is something particularly odious when a hospital refuses to examine or treat an injured person, denying access to their emergency department, and later to request for x-ray and other diagnostic tests; all because this is a work injury at the same hospital.

Egregious General Hospital had violated the most basic tenets of healthcare, as well as state and federal statutes.[1] No medical exam, no diagnostic tests, no preliminary treatment plan, no stabilization were provided by the emergency department at Egregious General, in violation of Emergency Medical Treatment and Active Labor Act (EMTALA). Over the ensuing 9 days, every effort to receive proper medical screening, diagnostic tests, and treatment was stonewalled by Egregious General; also in violation of EMTALA.

Much later, I would report the violations to Florida Agency for Healthcare Administration and to the federal Department of Health and Human Services, Centers for Medicare and Medicaid Services (CMS) to investigate for federal violations. CMS conducted an extensive investigation and after "careful review of the findings," determined that Egregious General Hospital had violated 42 CFR 489.24 in that it failed

"to provide for an appropriate medical screening, examination within the capability of the hospital's emergency department including ancillary services routinely available to the department to determine whether or not an emergency medical condition exists."

Further, they "determined that the deficiency is so serious that it constitutes an immediate threat to the health and safety of any individual who comes to the emergency department and requests examination or treatment for an emergency medical condition. Further, under 42 CFR 489.53, a hospital that violates the provisions of 42 CFR 489.24 is subject to termination of its provider agreement. Consequently, we plan to terminate _____Hospital's participation in the Medicare program."

The letter from CMS gave a termination date 19 days hence and an opportunity for the hospital to provide

"… credible evidence of correction of the deficiencies" in order to avoid termination. Several complaint control numbers were identified on the letter, leading me to conclude that CMS had followed up on other cases at Egregious General, in addition to my own.[2]

*   *   *   *

I thought now that I had gotten a diagnostic work-up and treatment would soon be started, that my ordeal would soon be over.

I dragged myself to work one third of the time over the next two months, working from home the rest of the time. There were days when I drove my standard shift car to work with my left leg so numb I couldn't feel the shift pedal, my left arm numb, and my neck in such terrible pain that I tried to memorize the position of the cars around me, so that I could avoid turning my head. I soon learned to turn my whole upper body in order to look to right or left.

I had also tried to get my hospital computer set-up ergonomically corrected. In my last effort, on May 11, 1999, my memorandum stated "Now it is urgent." It never was corrected.

"Pain doesn't count," the workers' compensation-assigned orthopedist would tell me. By this time the pain was constant, all-consuming, overriding everything I did, or tried to do, in my daily life.

After a total of 6 visits, plus 10 PT visits, this orthopedist pronounced me at maximum medical improvement (MMI) with 0% permanent impairment. MMI is a workers' compensation term that means this is the date beyond which no lasting improvement to an injury can be expected. It also signaled the end of eligibility for wage compensation (which had not been paid), and continuing medical treatment under workers' compensation.

Within two months, I had stopped working for Egregious General, unable to commute and get through a day without extreme pain. At the time, I did not realize that my injury was permanent. I thought once I got proper treatment, I would be well again. Indeed, once I was no longer putting in 12-hour days (work plus 3 hour commute) and resting a lot, I improved.

As the treatments stopped, the pain and immobility began again. Within a few months, I was back to square one, with as much pain as before. I returned to the orthopedist's office for treatment, and a referral for PT was made. But the workers' compensation insurer refused to authorize any treatment, and denied payment for the few treatments that had been provided. As we have seen, this is typical of what workers' compensation does. Treatment was as brief and cursory as they could make it, just enough to be temporarily palliative, but not effective to fix the problem.

What I also learned at that visit was that this was a recurring disorder that I am likely to have for the rest of my life. Why hadn't they told me that when I first came for treatment? Their explanation was that sometimes it goes away, and that it's hard to predict. But they were still labeling it as a muscle strain problem—very convenient for the cursory treatment that workers' compensation likes to provide.

While workers' compensation would no longer cover the injury, medical insurance wouldn't cover it either, because it was a work-related injury. Later, when I renewed my medical insurance, I had to sign a waiver that carved out anything having to do with my neck. Herein lies a typical workers' compensation conundrum: Workers' compensation insurance denies an injury is any longer their responsibility (the MMI determination), but health insurance will not be responsible for it because it's a work injury.

I sought the services of an attorney. I had truly believed that once I got proper treatment, I would be okay. I was not prepared for the journey I would endure over the next five years.

For two years all treatment, all therapy, all prescriptions, were denied by the insurer. My life became centered on how to get comfortable—a snatched few moments lying very still.

It was about this time, that I drew up a list of all the ways my injury was impairing my life: driving, as mentioned previously; exercising—I couldn't move my left arm from the shoulder when walking; couldn't hold the phone and write; couldn't lift a briefcase; get up from a chair without stumbling; when sitting at a desk there was no place comfortable to put my left arm; needed both hands just to hold a coffee cup; couldn't open a jar; push a grocery cart; wring a mop; lift a vacuum; couldn't be hugged around the neck; some sexual activities were now painful; I tired easily; endured pain and numbness every day, throughout the day. Even slight neck moves set off pain and sometimes a crunching sound. All of this added up to loss of stamina, loss of energy and inability to do anything for any extended period of time, and some things not at all. I saw my body deteriorating: the lack of exercise was causing a loss of muscle mass and turning to fat.

I intermittently used acupuncture and massage therapy, which I paid for myself. It all helped, at least briefly.

I also used a variety of aids, also without any reimbursement from the insurer. These included a soft neck brace, special orthopedic pillow, portable neck massager, massager that fits into chair or driver's seat, heat wraps away from home, and moist heat at home, ergonomically correct office chair, upright copy holders, headphones, and much later a gym program worked out by a trainer in consultation with my physician.

For a good part of my adult life, I had been a walker/jogger. Now I was unable to run, or even walk, for exercise. My legs were as strong as ever, but moving my arms was painful and exhausting. It was my son, Jamal Berkeley, who has a degree in product design, who created the solution that worked and that I still use today. I explained to Jamal how I couldn't support my left arm, and this was impairing my ability to walk for exercise. I had made poles cut from PVC pipe, with wheel chair hand-grips, and rigged wheels to pull the poles back. It didn't work. What Jamal devised was light-weight ski poles to almost shoulder level height, much higher than I would have used for skiing, but the height to which I had normally moved my arms when speed-walking. The ribbon loop on the left side is secured tightly and held just beyond my wrist, so that my hand is supported on the ledge of the handle grip. In this way, I don't have to grip the pole with my whole hand, only brace it against my thumb. The right side (my good side) is held normally. I also use ski poles with the treadmill, with the poles based on the floor. Jamal's concept was a life-saver: The modified ski poles allowed me to resume exercising in the same way I had done for decades before my injury. It is interesting to note that none of the workers' compensation practitioners with whom I spoke could suggest any solution other than a cane. A cane would, of course, compound the problem; it's the wrong height, has too much heft, and would require that I wrap my hand around it to hold it: all the opposite of what was needed.

Through the intervention of my attorney to re-open the MMI decision, the workers' compensation insurer agreed to an appointment with an orthopedic surgeon. He, in turn, recommended that I see a physiatrist, within the same practice. A physiatrist is a physician that specializes in non-invasive treatment, such as physical therapy and rehabilitation.

This was the physiatrist from hell. He pressed and probed about my head and neck, with neither warning nor explanation. Unbelievably, he later wrote in his notes, "Cervical compression produced no pain." According to Dr. N, most people have bulging discs and are asymptomatic, not feeling pain. But I am in pain, and my pain started suddenly and dramatically following a fall and tends to be on one side only. Therefore, how is his hypothesis applicable in my case, I asked him. He didn't answer. Dr. N was essentially denying that I was in pain. He didn't take kindly to questions about possible treatments, either. He was generally very patronizing, and he didn't listen: I told him I was highly allergic to lactose, explaining that it is sometimes used as a filler in medications. Despite this, he prescribed an anti-inflammatory drug which contains lactose. As it turned out, the insurer denied approval for the prescription; but not before stalling the pharmacist for two weeks and several phone calls while "the adjuster is still working on it." He also prescribed biofeedback, which the insurer denied as well. It took the better part of two months, several telephone calls and finally showing up in his office, in order to get a copy of my evaluation from this physiatrist.

By June, 2001, my attorney was successful in getting an appointment scheduled with an anesthesiologist and pain medicine physician. In retrospect, a referral to an anesthesiologist indicated a focus on relief of pain over amelioration of the problem. But the pain was certainly the overriding factor in my life.

Dr. D. diagnosed cervical disc displacement with radiculopathy. Cervical radiculopathy refers to irritation or pinching of a nerve, causing radiating pain down the shoulder, arms, or hands.[3] Up until now, despite the straightening evident on the 4/9/99 x-ray and the disc bulging evident on the 4/9/99 MRI, Egregious General's RN, the orthopedist and Dr. N, all had recorded the diagnosis as cervical strain/sprain. "Strain refers to an injury to a muscle. Sprain refers to a ligamentous injury."[4] But if the neck has lost its normal curve, then neck and shoulder muscles attempt to take over holding one's head up. Since these muscles were not created to do this, they tire and spasms occur. But the primary cause is the loss of the neck's normal curvature.[5] In other words, its disc and spine damage. The muscle strain is secondary.

Between September 1999, when my effort to return to treatment had been denied; and Dr. D's receiving approval for treatment, two years had elapsed. During this time, only evaluations were approved by the insurer. Egregious General and its appointed insurer had denied all treatment, all prescriptions for two years; while my pain, numbness, and immobility increased, and the permanent damage to my neck was becoming more severe. Had I been diagnosed and treated appropriately and completely in the beginning, would I have a permanent impairment now? I don't know the answer.

Dr. D prescribed a new MRI, which, when it was finally approved by the insurer, showed that the disc bulge at C4-C5 had not changed, but the C5-C6 bulge had now progressed to a herniated disc.

Getting the approval for a new MRI was quite an experience. When the MRI clinic called the insurer for approval after having faxed the prescription, they were told that I didn't need the MRI, that they were not going to approve it, and finally to "take the patient (me) off the scanner."

By this time, in addition to the pain, numbness and immobility, I was also experiencing pain in my left ear, a crunching sound when I turned my head, and difficulty swallowing at times. I had seen an ENT physician regarding the ear-ache, and was being treated symptomatically; but the treating physician assured me that the problem was not in the ear, that a more likely explanation was peripheral damage from the cervical spine injury.

Dr. D. prescribed physical therapy and a home exercise program. When the workers' compensation insurer approved PT, they wanted to send me to PT 50 miles away. It would take three months before PT would finally be approved. When the PT was approaching 6 weeks and physical exam was showing "residual cervical muscle spasticity as well as pain with flexion and extension," and "tender-ness to the paraspinals as well as over the cervical spinous processes." Dr. D. requested an additional six weeks of PT. This time it was approved. He also rec-ommended, and ultimately got approval for, a muscle stimulation and interferen-tial system machine. Physical therapy in combination with the muscle stimulation and interferential system machine was the most effective yet in reliev-ing pain. But over time, the effectiveness of this too dwindled.

Dr. D. was the only one of the four (25%) workers' compensation-approved doctors who evidenced a commitment to meeting the healthcare needs of the patient, but he still had to work within the parameters of what workers' compen-sation would approve.

Increasingly workers' compensation insurers are requesting Functional Capac-ity Evaluations (FCE). A FCE is supposedly a test of one's strength, flexibility and material handling ability. However, there is a gross lack of standardization, validity and reliability for the many FCEs available; with the professional back-ground of the evaluators widely variant, the testing time and components varying from less than two hours up to two full days, and equipment costs ranging from $400 all the way up to $75,000. It is not surprising then, that there is no research to justify the use of the FCE.[6]

The particular one that was used with me had this statement on the first page: "Throughout the Functional Capacity Evaluation there are indicators which determine if you are exaggerating your pain and there are tests which determine if you are giving your best effort. If you fail these, we are required to report this information." There followed a long list of possible descriptors for one's pain, i.e. sharp, dull, pulsing, sore, through tortuous, frightening, etc. On the next page was the body diagram, front and back. Despite the initial long list of descriptors, this page asked you to choose from only four words for your current pain: stab-bing, numbness, and two others, and to mark only those on the specific body parts for your pain at that moment. In other words, if you couldn't describe your current pain as one of those four, this test didn't want to know about it. But why not? What is the purpose of asking? What do only those four descriptors tell the evaluator that is important, while none of the other descriptors of one's pain from the previous page are considered important? Was this merely a badly designed evaluation instrument, or an expression of paranoia and suspicion of the patient, or both? The physical tests in this FCE were conducted over a few hours. But there was no FCE evaluator around the next day to ask me to identify my pain and soreness that had magnified, most likely because of the FCE.

Periodic depositions were requested by Egregious General's attorney. These lengthy inquisitions asked for everything: my birth certificate, copies of my passport, my license, my social security card; 10 years of medical history including names of all treating physicians, hospitals; household income, household expenses; income tax returns, convictions (there aren't any) and much more. These were really fishing expeditions to see if there was anything they could possibly use, or some way they could discredit either my claim for injury or myself.

Another battle was the concept that my work was "sedentary." According to Egregious General, I wasn't really impaired, because I didn't have to do anything physical in my work. As I explained to my attorney, my work involves my neck, head, shoulders, arms, hands 100% of the time; and those are the body parts impaired by my cervical spine injury.

How does an administrator move on the job? Reading, writing, researching, using the telephone, using the telephone while writing, leading meetings, carrying a briefcase, sitting, using the computer, driving; all involve head, shoulders, arms, hands.

I had tried to work twice following the injury, once in an administrative position, once not; but pain and numbness did me in within months both times. I had continually explored alternatives, consultant opportunities, part time positions, new careers, as well as positions commensurate with my experience; the latter in the hope that I could talk the corporation into a part time or consultant position once I got to the interview.

Midway through 2001, workers' compensation referred me to a contracted vocational evaluator for assistance in securing a new position. In line with their thinking on the sedentary nature of my work, they saw no reason why I shouldn't be able to hold a full time position. I dutifully followed up on all the vocational evaluator's leads, no matter how inappropriate, e.g. counseling, when I hadn't provided counseling, or direct patient care, in over twenty years and had no degrees in therapeutic counseling. At the same time, I was required to track all my work search efforts, whether ones I had developed, or those that had come from the evaluator; and submit that to the vocational evaluator on a regular basis.

I didn't learn, until much later, about the letter that Egregious General's attorney was sending to every organization on the job lead list I was submitting to the vocational evaluator. Fortunately, a considerate corporate attorney sent me a copy of the letter that one organization's executive search committee had received from Egregious General's attorney following receipt of my resume. The letter read as follows:

Attorneys at Law

Date

Re: Patrice Woeppel v._____Hospital
    Date of birth _____
    Social Security No. _____
    Our file No._____

Dear Sir or Madam:

Please be advised that we represent _____Hospital and (insurer), as a result of a workers' compensation claim filed by Claimant, Patrice Woeppel. We would appreciate your help in determining whether Ms. Patrice Woeppel has made any attempts to seek employment with your company. You can fax your response to us at _____ .

Your cooperation is greatly appreciated.

Sincerely,

I was livid. First, they were divulging my date of birth and social security number to those who have no right to know. But that was the least of it. They were sending me leads, and then sending a letter to those leads, that was basically the "Kiss of Death." Not only were they destroying their leads, which weren't worth much; but they were destroying my own, finely honed ones. I asked myself what planet these people were from.

After I called my attorney and demanded to have a copy of every letter they had sent, Egregious General's attorney apparently admitted they realized the job leads were not appropriate. They are reported to have said they didn't realize that the letters were so damaging. They promised not to send any more. I had already made up my mind; they were not getting any more information from me, upon which to send a letter.

Around the latter part of 2002, I began chiropractic treatment. I had requested chiropractic medicine several times, but the insurer would not hear of it. This time, I didn't ask. I knew I would pay for it myself.

I started with three chiropractic visits per week, then two times per week at the point where my range of motion had increased and function improved. Once I was feeling well again, treatments continued at one visit per week to prevent soft tissue contracture and to maintain my range of motion. Maintaining proper intersegmental mobility limits degeneration and lessens the possibility that my pain and impairments will return. Proper bio-mechanics in my spine also allow my muscles to function properly, so that I have improved strength and functionality.[7]

I continue chiropractic treatments to maintain optimal function. My head no longer feels like its too heavy for my neck. I am able to move, drive, walk, work on the computer, turn my head when I choose. I am so grateful to my chiropractor, Dr. Andrew Charni: He's given me my life back.

I was more fortunate than many: I was 60 years old when the accident occurred, thus I was nearing the end of my career. The injury and its aftermath, of course, torpedoed it.

My savings and my retirement fund were used for living expenses; for the medical insurance that carved out my neck; for treatment for my neck; and to pay off the balance on student loans, most for my doctorate. At age 62 years, three years earlier than planned, I began receiving Social Security.

It was five years before my workers' compensation case was settled. In that time, wage compensation had not been paid. The settlement agreement does not allow me to divulge the actual amount of the settlement: I can say that the amount of the settlement was only a minute percentage of the actual cost of my occupational injury.

The un-reimbursed cost of my occupational injury, after settlement, has reached almost $700,000. Included in that calculation are lost wages and benefits; additional cost of health insurance to age 65 years; uncovered medical treatment and out-of-pocket injury expenses; and the over-all reduction in Social Security benefits, projected over my expected lifespan, based on having taken Social Security at age 62, minus the three years of early benefits.

My husband, Don, and my son, Jamal, have been my emotional support throughout this ordeal.

Had I been anyone who just walked into Egregious General Hospital and slipped on that dirty floor, both the outcome and the events along the way would have been quite different. I would have had the leverage of a potential civil suit. But as an injured employee, all I had was workers' compensation, in which the scales of justice are heavily weighted against the injured worker.

My own experience with this sham of a system that devastates the health, the lives, the assets of injured and ill workers and their families throughout the country, on a daily basis, led me to research, and ultimately, to write this book.

References:

1. Florida Statutes, 395.1041; and federal statutes 42 CFR 489.20, 42 CFR 489.24.

2. Letter, Department of Health and Human Services, Centers for Medicare and Medicaid Services to _____CEO,_____Hospital, dated December 4, 2003.

3. Cervical Radiculopathy, www.emedx.com/emedx/diagnosis-information/

4. Malanga, G.A. Cervical Spine Sprain/Strain Injuries. Emedicine from WebMD, 4/15/05,www.emedicine.com/SPORTS/topic24.htm.

5. Chiropractic Wellness and Fitness Magazine, V. 3, Issue 2, p. 39.

6. King, P.M., Tuckwell, N., Barrett, T.E. A critical review of functional capacity evaluations. Physical Therapy, Vol. 78, No. 8, August 1998, pp. 852–866.

7. Information from Dr. Andrew Charni, Family Chiropractic Center, Margate, FL, November, 2005.

▼

CHAPTER 10

MARITIME DANGERS AND DIFFERENCES

Maritime workers have historically been covered under the Jones Act, a federal law.[1] Unlike workers' compensation, an injured or occupationally exposed merchant marine has a right to choose their treating physician, can be compensated for pain and suffering and, most significantly, can sue under the Jones Act.

The Jones Act "provides heightened legal protections to seamen because of their exposure to the perils of the sea ..."[2] The owners of a vessel have an absolute responsibility to "provide a seaworthy vessel ... a safe place to live and work. Even where a vessel is seaworthy when it leaves shore, it can become unseaworthy on the basis of dangers which arise or are created during its voyage."[3] The Jones Act permits injured seamen to recover lost earnings and future earnings while disabled, medical treatment with practitioners of their own choosing, vocational retraining, living expenses while recovering, as well as compensation for pain and suffering. Actions may be brought in state or federal courts.[4]

*   *   *   *

James Bowles met me in the driveway of his neatly kept mobile home in this rural area of coastal northeast Florida. He walked swiftly, with purpose. But now, it's all between bouts of coughing and losing his breath, and exhaustion from the energy these episodes take from his body. Jim Bowles is 46 years old.

On January 3, 1997, Jim Bowles whole life changed forever. He knows what he is enduring now. What he doesn't know is what's in store for him. Cancer is likely, he explains; but no one can predict what results from inhalation of propylene oxide, at the extraordinarily high levels to which he believes he was exposed.

The US Coast Guard has the statutory authority "over the safety and health of seamen on board vessels which are inspected and certificated by the US Coast Guard, i.e. 'inspected' vessels."[5]

The chemical spill by the vessel should have been reported to the United States Coast Guard National Response Center (NRC) as required by the Federal Water Pollution Control Act.[6] A spillage of more than 119 gallons, which this is believed to have been, would have been required to be reported.[7] But there was no report that I could find. Under Emergency Planning and Community Right to Know Act of 1986, Section 313, "releases of more than 100 lbs/24 hours of propylene oxide into the air, water, and land must be reported annually and entered into the Toxic Release Inventory (TRI)."[8] But again, there was no report.

Illegal spillage or disposal of waste and/or pollutants into coastal waters clearly exists. One recent investigation reported one company's oil-contaminated waste dumped from nine ships off six US ports.[9] Despite the unknowns, what is recorded, shows Texas to have the largest amount of propylene oxide released of any state in 1998, at 268,455 pounds.[10]

There were, however, other validations for the events of January 3, 1997, such as the settlement agreement with the harbor tug company, dated August 21, 1997.

Propylene oxide, the chemical to which Jim Bowles was exposed, is a "volatile, clear, colorless, extremely flammable liquid with an ether-like odor. It is soluble in water ..., (and) may form explosive mixtures with air."[11] Propylene oxide "may reasonably be anticipated to be a human carcinogen, according to the Sixth Annual Report on Carcinogens, published by the National Toxicology Program, US Department of Health and Human Services. It is also classified as a carcinogen in EPA's Toxic Release Inventory (TRI)."[12]

Propylene oxide is a "severe irritant of the eyes and respiratory tract. It is a mild central nervous system depressant. Acute inhalation exposure may result in ... headache, motor weakness, incoordination, and coma. Respiratory symptoms, which may be delayed, include coughing, difficulty in breathing, and pulmonary edema. Neurotoxicity has been reported in animal studies."[13] "Dermal (skin) contact, even with dilute solutions, has caused skin irritation and necrosis in humans."[14] National Institute of Occupational Safety and Health's immediately dangerous to life or health (IDLH) concentration is 950 mg./m3. This is the upper exposure limit "to ensure that a worker can escape from an exposure condition that is likely to cause death or immediate or delayed permanent adverse health effects...."[15] The concentration to which Jim Bowles estimates he was exposed could be vastly higher than the IDLH.

Jim Bowles was 37 years old at the time. Then, he was in "great health, great physical condition. I always did hard labor, was able to (work on) houses, carry shingles, pull lines, any type of labor I wanted to do."

Bowles had joined the Seafarers International Union back in the late 80s. "I worked in Hawaii on my first ship, a cruise ship, as a deckhand. As the jobs came up, I transferred down to the Engine Room, where I was told I'd be best because of my mechanical knowledge and ability, and from that point on, I always worked engine room. I sailed deep sea for several years on numerous ships; tankers, container ships, freighters, doing my job. I started out as basically a wiper in the Engine Room, and worked my way up the ranks to one of the top ratings of the unlicensed people, which is QMED, Engine Department" (Qualified Member of the Engineering Department). That is the highest unlicensed position one can go, Jim explains. "I was studying for my Third Engineer's license at the time I got out of deep sea."

Jim got married and his wife didn't want him leaving home for months at a time. He transferred from deep sea to inland and took up sailing on the harbor tugs, which is where his exposure to propylene oxide occurred. Jim explains that a harbor tug is used for docking and undocking large ships. When the deep-sea ships come into a port, because of their size, they are not able to steer the channels, and they can't stop. From 1 to 6 harbor tugs are required, depending on the size of the ship, to throw out lines up onto the ship, tie them off to the tugs, and then use the tugs for the steering and the brakes.

"I was living outside Freeport, where chemical tankers and oil tankers came in to drop off oil, to be pumped through the pipe lines, and into the national oil reserves, and into the sulfur domes in Freeport, Texas."

I had to work my way up the list from the bottom entry position. At the point where I was injured, I was working as a Chief Engineer. I worked as a Relief Chief when one of the other Chiefs wasn't around. That's the top of the Engine Room ladder.

These Freeport tugs were a one-crew tugboat. We were supposed to do repairs and maintenance on the tugs from 6:00 AM till noon. And if we had no ships to bring in, we had the rest of the day off. Saturday, Sunday, holidays off and anything after noon, unless a ship comes in. So it sounds wonderful. But all the ships come in the afternoon, so we'd be bringing ships in from noon until 6 in the morning, and you would have a beeper attached to you. A normal worker had an hour call-back to get to the ship, and be on the ship. But, working in the engine room, I had to be there in 45 minutes; so I could get on the ship, fire up the engine, let it warm up before we were due to work..... I was on call 24 hours a day, 7 days a week. I don't believe I remember, ever going more than a day where a ship didn't come in, after the hours that we were off work. Then they decided for safety reasons that they had to have one person stay on the boat and do a watch. A rotation among the four people meant that once every 4 days you had to stay on the ship; ... and monitor your tugboat, and the engine room in the other tugboat every hour, to make sure nothing was on fire, or burning, or nothing low on oil.

"On January 2, 1997, we were called to dock a tanker carrying propylene oxide. We brought the ship in about 10 or 11 pm on January 2nd. The ship had come from overseas, somewhere up North, where the water temperature was quite cold. In the Texas Gulf stream, the water temperature is considerably warmer. The cargo had expanded and the tanks were over-pressurized, to where they couldn't hook up to the company's dock and unload the ship, without bending the tanks. So we were called back to undock the ship, and set the ship out to sea. They were required to go 25 miles out into international waters, where they would dump the overflow in the tanks, let the tanks and any spillage go right over into the ocean." Jim explains that it's standard procedure to dump anything into international water. He says they didn't go out that far. "A tanker that can reach top speed of about 16 knots loaded, (about 22 miles an hour) would take approximately several hours to get up to speed, and then whatever time it took to get 25 miles out to sea; the time it takes to unpressurize or unload the tanks out there, and then several hours to get it fired up, turned around, and back into port. The ship, they believe, went out 3 miles to the sea buoy, and dumped out right off the coast. The next day, Jim reports, there were a number of people suffering respiratory problems on Surfside Island and Quintana Island, islands right off the Intercoastal Waterway, on the Gulf.

"We undocked the ship and within a very short period of time, ... got our tug back to the dock, shut the engines down, ... I hopped in my car. I lived approximately a 45-minute drive away at the time. I was about half way home when my beeper went off saying we had to bring the ship back in. So there was no way possible that ship could have made 25 miles to sea. A nuclear aircraft carrier couldn't have made it in that time," says Jim Bowles.

"They brought the ship back in, and we were sent out to bring it in the channel. As the ship approached us, and as we got our lines up on board, the tugs were overwhelmed with strong benzene-smelling, acetone-smelling fumes. We got our lines attached. We were using our tug to slow the ship down. I was relief Chief Engineer at the time, because the regular chief was off, along with the regular captain. So we had a relief captain, and I was sitting in for chief engineer, and we had two deck hands on the deck working the lines. During a normal job, I, as chief, sit up in the wheelhouse with the captain; and if there is trouble with the winches or any of the engine stuff, I have to run down and take care of the engine. And if the guys on the deck have problems, if we have to put a stern line, which is an extra line off the back of the tugboat to the ship, I help them put the line out or bring it back in. For two men to bring in 4 feet of line, approx. 300 feet long; once it hits the water, it's approximately 100 lbs. per foot in weight. So you have to drag this heavy stuff up on deck. I assist them pulling this line up, so we don't get it caught in the prop of the ship.... If everything is going smoothly, I sit up there with the captain and the pilot of the ship, as we talk back and forth on the radio. But if anything happens, my job is to maintain the mechanical equipment on the ship, make sure everything is working properly. A lot of chiefs won't, but if something happens on the deck, and the guys are having trouble, three people deal with it a lot easier than two." Jim sees it as common courtesy that any good chief would help as much as he could.

"As we brought this ship in, we were all instantly hit with this very strong, obnoxious smelling stuff. Instantly our eyes were burning, our throats got real rough, almost like an instant sore throat, very aggravated, extreme trouble breathing, tightness in the chest. The best I could explain it, it would be like opening a bucket of ammonia in a small room and letting the ammonia just tear you up. It was an extremely toxic smelling chemical, and everyone on board instantly reacted." Jim described it as "so strong, it was like sticking your head in a bucket full of a high powered chemical with no air around it. It was solid fumes. They estimated that we were exposed to probably 35,000–55,000–75,000 ppm," or several hundred times the permissible exposure limit. Jim explained that according to the MSDS, propylene oxide's maximum exposure is 100 ppm that a person can be subject to while working. Indeed, that is also OSHA's permissible exposure limit.[16]

"Several people on the tanker suffered inhalation injuries too. When I was in the doctor's office for my injuries, I noticed there were two guys from the tanker, also having respiratory trouble. I think they sent them back to wherever they were sailing from (sic).

"The two deckhands, who were out on the deck handling the lines while I was up in the wheel house, were overcome by propylene oxide so strongly, that they were passing out on the deck. I had to go down and assist, and literally carry one guy up to the wheelhouse.

"Propylene oxide is an inert gas which is heavier than oxygen. The fumes come right off the ship, which is approximately 40 feet up in the air from the main deck to the water, and the harbor tug sits much lower than that. The fumes rolled right over the deck and engulfed our whole harbor tug. So anyone near the water was getting the strongest amount of it, and the safest place on the ship was all the way up in the wheelhouse, the top point on the harbor tug.

"Under normal conditions, with a spill like this, we would take a fire ax, chop our lines and release the tug, and get the heck away from it. We had a relief captain on board, and he was afraid to give the command to cut the lines; so we stayed in this chemical for well over an hour, until we got the boat docked at the dock.

"If we would have cut it loose, coming into the channel, there's no stopping this thing (the tanker); it's loaded with flammables. If it would have caught the plant on fire, or crashed into the fuel docks on the other side, the Gulf of Mexico would probably (extend) out to Oklahoma or California now." Jim explained that not only is propylene oxide flammable, but all the other chemicals in the area are flammable too. "The refineries are making gasoline; and nearby is the nation's storage for crude oil, the sulfur domes containing billions of barrels. If this area caught fire, all of Texas would be gone, just from the explosions in the pipelines. So they didn't want to cut the boat loose, or run it aground, or try not to have the problem."

There are no less than 27 chemical plants and facilities containing 131 different chemicals in the Freeport area.[17] The worst industrial accident in US history reportedly occurred just a few miles away in 1947, when a fire aboard a ship triggered a massive explosion "that killed 576 people and left fires burning" for days.[18]

"But at the same time, we were all injured. We were at the downwind side, so our tug had a larger exposure than the other tug involved in docking this vessel. All four members of the other tug were also affected.

"All eight of us were sent to the emergency room that night, Friday, January 3, 1997, and we were all in the ER late into the night. I believe it was early Saturday morning, 1 or 2 am, when the emergency room doctors said they had no idea what propylene oxide was, or what kind of damage it could cause; but from what they could understand it was just an irritant, and that we should follow up with our regular doctors, and that there was nothing they could do other than to let us go.

"Maybe we went back to work, because I have on my calendar that I was put out of work on January 8th, as unfit for duty. That was when the doctor declared me unfit for duty because of the problems with my lungs. After several visits, he was unable to find any help or anything to do for me, so he referred me to a doctor in Houston, a respiratory expert. I was under his care probably from about January 12, 1997 through July 23, 1997. I had to beg the doctor to release me for duty because of my financial situation. I reported back, had to go in and take a physical from the union, so they could declare me fit for duty after the doctor gave the approval.... I reported back to work with a clinic card saying I had passed my physical July 28,1997. They put me back on board (another tug) on Tuesday, July 29, 1997, at midnight."

This "was an older tug with a different type of diesel engine in it. It smoked extremely badly. Now I'm suffering from severe migraines, all kinds of respiratory problems. After being off work for seven months under doctor's care, I still have chemical airways disease. My lungs were burnt from the chemical. My capacity to transfer oxygen into the blood stream now is greatly reduced. Ever since the injury I have blue feet because I don't get enough oxygen in the blood. When they monitor the oxygen, it runs from a 0 to 10, with 10 being the greatest amount and 0 being dead. I believe mine was a 1.5 to a 2.0, the last time I was tested—very low on the scale." Jim's medical records document that his obstructive airways disease has been growing progressively worse over time.[19]

"Before the year was out, I was off the ship again, unfit for duty and unable to work; and I had to leave my job at that time. At this point, I was not fully vested with the union. I didn't have my ten years of sea time in with the union; therefore when I was disabled from work, I was not entitled to any type of pension, any type of health care or any type of (benefit) whatsoever.... The only thing I had left to do was to go back and sue the tanker company and/or the other company for my injuries."

While he had been out on sick leave, Jim received a percentage of his wages, but "not enough for me to live on with my family. I was used to making anywhere from $1500–2000 every two weeks and I believe it was well under half of that. On August 21, 1997, I signed a release form, because they weren't responsible for my injury; and I was being given the balance of my wages for that period of time, in addition to what they had already given me. So I released them." Jim adds, "This was not any admission of liability or anything else. It wasn't their problem." In Jim's view, the fault was a combination of the company that sent the ship out to dump the chemicals and the shipping company not doing it properly.

Jim's settlement does not allow him to be in contact with any of his fellow workers. This is a common restriction in such settlements. Jim believes he was the most seriously injured of those who were exposed that day. "I was the one on the ship that got the most exposure, and I was also a lot older than the young kids on board. They were about 21 years old. I had previous problems with slight asthma before I went away. The fact that I was 16 years older and had worked in a lot of industrial areas; power plants, coal fire, fossil fuel generation, gasification plants in the past, would have had some influence on the condition of my lungs at that time also.

"After receiving the money to survive from (harbor tug company), I was still unfit for duty; and the only way to win any compensation in an injury is to go back and sue the companies involved."

Longshoremen's and Harbor Workers' Compensation Act, provides that the acceptance of the workers' compensation settlement from his employer would have eliminated his ability to recover from the shipping company or others, unless he initiated the process within six months. Otherwise, any proceeds would be required to be turned over to the tug company, with which he had settled.[20] Within the required six months, Jim started the lawsuit, using a firm of maritime attorneys in Houston.

Because he is a maritime attorney, Jim explains, he doesn't have to limit his fee as a workers compensation attorney would have to do. The firm was also able to provide Jim with a monthly check to cover living expenses, while awaiting settlement of the claim.

Jim moved from Freeport, Texas to Florida. "Because of the respiratory problems I have, the doctor had recommended that I get out of the Houston and Freeport area. All the refineries, chemical plants, the poor quality of the air, the pollution from the cars; it's one of the most polluted places in the United States. He recommended moving to either Arizona or Florida; because the dry climate of Arizona, or the salt air of Florida, would be good for my respiratory system. Having family in Florida, I picked Florida." Jim selected an almost rural area with no industry and not a lot of roads, thus reducing his exposure to both industrial and vehicle emissions.

After he moved to Florida, "I had several trips back and forth to Houston for doctor visits, pulmonary function tests, other lung tests."

"I suffer from constant chest pain and sore lungs, most days severe to extreme pain. On a good day; two days a week, sometimes as many as three, or as few as none, the pain is not quite as severe. But it is always there.

"I have headaches and migraines almost every day, from annoying to so severe, at times I lose my eyesight …

"I have involuntary respiratory attacks, where I vomit up lung discharge; and these are extremely painful to the chest and the lungs. And if I have a headache at the time, the headache gets real severe and my head starts throbbing in my temples. I've had these respiratory attacks so bad that I've actually blacked out. Fortunately, I can always feel them coming on before they hit. If I'm driving a car, I know I'm going to have one and it gives me time to get off the road; so that I can start coughing up on the side somewhere, rather than being in the car, and possibly black out at the wheel."

Jim also has bouts of diarrhea, for which he takes preventative and over the counter medications to control and allow him to leave the house.

"I have joint pain that moves around at will, or in between joints. It will start out like a bruise and intensify so severely, that it feels like I've had a severed limb. The extreme pain sometimes will last 2 to 6 hours and then stop. Sometimes it will last 10 minutes. I went 17 days with one of them.

"I have limbs that get numb or fall asleep whenever I lay down; and I get severe muscle cramps, Charlie horses, in the legs, mostly when I'm laying down.

"I've been suffering ever since this injury, hot and cold sweats—bad. And, I'll be freezing at one moment and the next minute, I'll be soaked like you dumped a bucket of water over me, completely saturated.

"I get involuntary muscle spasms, that will start twitching. Fortunately those have gotten better since my original injury. I don't have the muscle spasms very often. Every now and then I'll still get one, like an eye will start twitching and I can't control it, or something in an arm or a leg."

"I experience pain all the time, 24 hours a day, but certain things make it worse: Any physical exertion, like walking, running, trying to move fast, lifting anything; causes me to lose my breath, my chest pain will go up. Certain substances; like perfumes, car exhaust, diesel exhaust, various household cleaners, gasoline fumes can cause me to go into extreme attack. I can't go into air-conditioned spaces because they cause me to have extreme pain in my lungs. Temperatures under 50 degrees Fahrenheit cause extreme chest pain and tightness, that make it hard to breathe."

When the case went to court and was settled; Jim's attorneys were reimbursed for the money they had advanced him for living expenses each month, and medical bills, and other expenses they had paid, and their fee. "They gave me the small remaining difference back and that was my lifetime settlement, to live on the rest of my life; which for an average person might last a year, maybe two, if they stretched it."

Jim describes his present financial situation as "very difficult." He has a renter who pays him $100 a week. "With the settlement that I got, I purchased the trailer, so I don't have any mortgage. I live on a very minimal amount, basically $400 a month."

Jim had to remove the carpeting and install wooden floors throughout his mobile home, in order to reduce the allergens and irritants in the air. Outside, leftover wooden slats became rustic reindeer planters and lawn chairs. Jim also saves old car parts, to recycle as needed. Nothing is wasted.

He had to sell his truck, but he uses his brother's aged sports car in return for keeping it in good shape. Jim opened the hood, exposing an immaculate engine and explained the latest repair. He had been a licensed mechanic before becoming a merchant marine. In high school, he had won a scholarship to the General Motors Goodwrench Technician training program. Jim completed the five-year course in three years and was the first student to graduate. Jim's work on the car of a Denver-based corporate quality assurance manager so impressed the manager, that it launched Jim's next career as an inspector in a nuclear power plant; and ultimately he became the site quality assurance manager at a nuclear power plant.

Jim has a small workshop area, very neat and well organized. Pulling the heavier drawers open caused evident pain.

Jim Bowles is a resourceful man, e.g., creating a well for water, so he doesn't have to rely totally on city water; or utilizing the water one runs until it gets to temperature, by feeding it to the washing machine for the next load of clothes.

He creatively uses everything he has, to the fullest extent of which he is capable. Despite his illness, Jim keeps active doing the things he loves. But now everything is paced and framed by what his illness allows him to do, and for how long. Jim says he is most comfortable when he is moving, but can't do anything for any length of time. He is only able to sleep a few hours at a time. Sometimes he sleeps in a recliner, because it is most difficult to breathe when he is lying down. Eight to nine am is his best time, he tells me. This day, he has been up since 4 a.m., and takes short breaks during our interview.

He still donates time to charities, and he donates blood. He says this not only benefits the blood drive, but it helps to replenish and renew his own, thus helping him to stay healthy.

Jim had traveled all over the world through his work. He had collected many items from different places. Many he has sold, as he needed money to survive; till now he has only a few left.

Jim Bowles has applied twice for Social Security Disability—unsuccessfully. He is in the process of applying again. Many injured and ill, disabled workers find that it is almost impossible to receive Social Security Disability benefits; without persevering through several attempts and hiring an attorney. This is a sad commentary on a system that should be a safety net for the disabled.

In Houston, Jim had used acupuncture to good effect. His problem now is he can't afford medical care. He has no coverage. He has been prescribed a slew of medicines, to assist with the many problems resulting from the toxic exposure he endured. Jim shows me a packet, about two inches thick, of prescriptions that he can't afford to fill. And some meds, that he did fill, are lining a large dresser top in his bedroom. He is unable to tolerate many of these drugs, and thus he doesn't take them. He knows he is a high risk for cancer.

"The lungs are an amazing thing. There's a possibility, they claim, a slight possibility, that my lungs might improve over time; and that I might be able to get my lungs back again, where I could come back into work. They also told me that propylene oxide is a highly carcinogenic chemical, and that they also expected me to come down with severe cancer before May of 1997.... Now it's 2005, but anyway.... I'm still alive."

Jim Bowles, and the other merchant mariners on board the harbor tugs that day, put their health on the line. In so doing, they avoided a massive explosion and disaster of monumental magnitude.

Jim's life has been altered dramatically by his illness: his health has deteriorated, he has been impoverished, he faces an uncertain future, and possibly an early death from cancer.

As a maritime worker, however, Jim Bowles had two advantages that workers' compensation victims generally do not have: 1) Jim Bowles was able to obtain wage compensation and medical treatment, without delay, from his employer; because his right to sue was not denied. His medical treatment was of his own choosing, and his employer settled quickly. 2) An attorney could afford to take his case, and was able to provide a monthly stipend to him, and needed diagnostic and medical treatment; until the final settlement with the companies responsible for his toxic exposure and subsequent illness.

None of this negates the extreme difficulty caused by the lack of sufficient data on many occupational toxic exposures and resultant diseases. The right to sue is essential and necessary, but insufficient by itself. Jim Bowles' final settlement leaves him impoverished and without the medical treatment that he needs.

The long latency period, combined with the lack of sufficient disease data on occupational toxic chemical exposures and resultant diseases, in general, is a continuing problem. The lack of specific knowledge of the toxin, propylene oxide, the amount to which Jim Bowles was exposed, and its long-term consequences; certainly would have hampered both his ability to receive adequate medical treatment, and the outcome of his legal case.

References:

1. 46 U.S.C. 688.

2. Chandris, Inc. v Latsis, Certiorari to the United States Court of Appeals, Second Circuit, decided June 14, 1995, www.admiraltylawguide.com/supct/Chandris.htm.

3. Maritime Injury Law and the Jones Act, by Aaron Larson, October 2003, cited at www.expertlaw.com/library/pubarticles/Personal_Injury/jones_act.html.

4. Injury at Sea, www.maritimeinjury.com.

5. CPL 02–01–020 OSHA/US Coast Guard Authority Over Vessels.

6. 40 CFR 116: Designation of Hazardous Substances.

7. 49 CFR 171.15: Immediate notice of certain hazardous materials incidents; and 49 CFR 171.16: Detailed hazardous materials incident reports.

8. Propylene Oxide Chemical Backgrounder, at www.nsc.org/library/chemical/propyleO.htm.

9. For fouling region's waters, shipping firm assessed record fine. Daily Hampshire Gazette, p. A7, 12/20/06.

10. Propylene Oxide Chemical Backgrounder, at www.nsc.org/library/chemical/propyleO.htm.

11. Propylene oxide, CAS No. 75–56–9, at http://ehp.niehs.nih.gov/roc/tenth/profiles/s155prop.pdf.

12. Propylene Oxide Chemical Backgrounder, at www.nsc.org/library/chemical/propyleO.htm.

13. US EPA, 1994a, cited in Propylene Oxide, CAS Registry Number 75–56–9, www.arb.ca.gov/toxics/tac/factshts/propylox.pdf.

14. Propylene oxide, hazard summary, cited from EPA's Integrated Risk Information System,(IRIS) at www.epa.gov/cgi-bin/epaprintonly.cgi.

15. Propylene oxide, NIOSH IDLH, page 4 at www.epa.gov/cgi-bin/epaprintonly.cgi.

16. US Department of Labor, OSHA 29 CFR 1910.1000, Subpart Z, Table Z-1 Limits for Air Contaminants, at www.osha.gov.

17. Dozens of Chemical Plants lay in Rita's path. National Environmental Trust, September 26, 2005, www.net.org/health/Texas_toxics.vtml.

18. Easton, Pam. Refinery Blast Kills 14:More than 100 injured in oil plant blaze in Texas. The Associated Press, cited in South Florida Sun-Sentinel, p. 3A, 3/24/05.

19. Medical records for James Bowles, 1997–2003.

20. 33 USC 933 (b) cited in Rodriguez v. Compass Shipping Co., Ltd., Et Al, 451 US 596, Certiorari to the US Court of Appeals for the second circuit, decided May 18, 1981, cited in www.admiraltylawguide.com.

CHAPTER 11

▼

SAVING THE ENVIRONMENT DEMANDS RIGOROUS REGULATION OF INDUSTRIAL TOXINS

One cannot see the repeated toxic chemical exposure of workers without also seeing its fallout on communities. There are long-term consequences to this devastation for individual workers, their families, communities, for society and ultimately, for our planet. Many of the health and environmental impacts are not even fully known yet—none of it good. Why is it that the connection between the tens of millions of workers exposed to toxic metals, or neurotoxic solvents, and other toxic organic compounds, other carcinogens, is not identified as the monumental environmental disaster that it is? Like the canary in the coal mine, worker toxic exposures are the harbinger to environmental devastation. What follows are a few examples.

In Bhopal, India, in 1984, a gas leak at a Union Carbide plant reportedly spewed 40 tons of lethal methyl isocyanate into the surrounding community, resulting in an immediate 3,500 deaths. Twenty years later, it appears the death toll from the disaster stands at 15,000–22,000; and another 800,000 persons still suffering the impact.[1] For many Americans, this tragedy was far away and quickly forgotten.

In Libby, Montana, an entire community had reportedly been contaminated with asbestos-containing vermiculite by the W. R. Grace Company, resulting in the deaths of over four hundred residents, and another twelve hundred with asbestos-related diseases.[2] Since the 1920's vermiculite ore had been mined near Libby. EPA did not place emphasis on asbestos and vermiculite contamination "due to funding constraints and competing priorities." EPA acknowledges that it did not "sufficiently address asbestos-contaminated vermiculite at Libby" and admits it "did not issue regulations under air and toxic substances statutes that could have protected Libby citizens from exposure to asbestos-contaminated vermiculite."[3]

In western New York State, the notorious Love Canal toxic waste site took 21 years and $400 million to clean up, evacuated hundreds of families from their homes; and even today, it is said that tons of toxic material are still buried there.[4]

Children and the elderly are the most vulnerable to the environmental pollution caused by industry.

The kids in Ashland, Massachusetts back in the mid-1960s through early 1980s reportedly thought it was pretty cool to be wading through blue, red, and purple water; near a dye manufacturing plant.[5] The town of less than 15,000 residents had been home over the years to various dye manufacturers beginning in 1917. The final consolidation, operated as the Nyanza Company, began in 1965 until it's closing in 1978. It was later identified as an EPA Superfund site, on which EPA has spent $46 million to clean up.[6] The discharge of volatile organic compounds (VOCs), dye manufacturing compounds and metals into unlined lagoons at the site, and nearby brooks, and wetlands had earned it a place on EPA's National Priority List.

In 2006, Massachusetts Department of Public Health (MDPH) completed a retrospective cohort study in which they interviewed 1387 residents (response rate of 67.5%), who had lived in the area when they were ages 10–18 years of age, between 1965 and 1985. They discovered 73 cancer cases, of which 40 could be medically confirmed. Fully 25 of the cases (34%) were rare forms of cancer. The study found a two to three times greater risk of cancer for those who had water contact, especially wading or swimming, in two areas of the site, one of which included waste lagoons. They also found that the risks were greatly increased for those who had had both water contact in those specific areas; and who also had a family history of cancer, which MDPH concludes is suggestive of gene-environment interaction.[7]

In the small town of Herculaneum, Missouri, the Doe Run Company had operated a lead smelter since the 1960's that employed many of the townspeople. Occasionally the company reportedly would test the tap water, or provide grass seed to cover the bare spots in folks' yards, or pay for the repair of the corroded paint on residents' cars caused by an acid plume. Leaded gasoline and leaded paint were banned in the 1970s because of the well-documented severe and irreversible damage to brain and central nervous system that occurs from exposure to lead, particularly in children. But even after years of EPA remediation efforts, the thick black dust along Herculaneum's streets en route to the smelter was still identified to be almost 30% over clean air standard limits for lead.[8] In 2005, blood lead levels tested in children under 6 years of age showed that 23% of the children living 1 to 1 $^{1/4}$ mile from the smelter had high blood-lead levels, increasing to 56% of the children living within a quarter mile of the smelter.[9]

According to a report issued by the Environmental Working Group, unborn babies are being exposed to "a stew of chemicals, including mercury, gasoline byproducts and pesticides." Tests were performed on ten samples of umbilical cord blood, finding a shocking 287 contaminants present; including pesticides, fire retardants, mercury, and others. Of the 287 chemicals present, 180 are known carcinogens, 208 known to cause abnormal development and birth defects in animal testing, and 217 that are toxic to brain and nervous system. Clearly we need to strengthen controls on industrial chemicals.[10]

In 2003, in Huntington Beach, California, a four-year old boy died of a brain tumor. The incidence of brain tumors in children is less than four children for the whole country.[11] But he was one of four children living within a two-mile radius of each other in Huntington Beach, California reported to have died of a brain tumor. The oldest was eleven years of age. It appeared that the upscale homes had been constructed on what had once been a dumping ground for the country's large gasoline companies.[12]

In Norco, Louisiana, a retired teacher watched her family and neighbors fall ill and die. Her African-American neighborhood sat between two petrochemical plants. She set up a webcam showing the illegal venting of toxic chemicals from the plant, and then proceeded to install her own atmospheric monitors. She is said to have personally traveled to corporate headquarters in the EU, inviting executives to come and see for themselves the air pollution they were causing. Eventually, the company is reported to have agreed to expend $20 million in emission reduction, and buying out the homeowners living near the plants.[13]

Nitrates in drinking water, found in agricultural areas, are associated with increased risk for non-Hodgkins lymphoma. Arsenic is used in pesticides and mining. In the drinking water, it causes skin and bladder cancers.[14]

Outbreaks of a lethal strain of E. coli after ingestion of raw vegetables are a recent phenomenon—a byproduct of the feedlot method of meat production, and centralized industrial agriculture methods of processing and packing. Animals no longer graze in open fields chewing on grass, which had a natural resistance to E. coli. Instead, feedlot cattle are confined in very small space; eating grains, which are an ideal host for E. coli; standing in their own excrement, also a natural habitat for E. coli; and being fed large doses of antibiotics, in a wrong-headed effort to counteract the effects of all of the above.[15] According to the Centers for Disease Control and Prevention, the American food supply now sickens 76 million persons each year, hospitalizing 300,000 of them and killing 5,000.[16]

There appears to be a disastrous disconnect between the Department of Agriculture, which regulates meat and dairy products and the Food and Drug Administration (FDA), the latter which regulates fruits and vegetables—well, sort of. In actuality, only the processing plants are within FDA's purview.[17] And the pathogenic ooze that comes from the feedlots, as well as local sewers, is being spread on farmland, without regulatory restraint of the full chemical content.[18]

For the members of one National Guard unit, their exposure in Iraq would come home to their families. Depleted Uranium (DU) is radioactive and toxic. It is said to damage bones, lungs, kidneys; cause loss of bowel and bladder control; cause fatigue; brain deterioration; and many forms of cancer. It is also reported to cause genetic damage and cross the placenta, causing damage to bone, muscle, cardiovascular, and central nervous system of the fetus.[19] Nine members from the Orangeburg, New York-based unit are said to have discovered that they were experiencing similar and debilitating symptoms. The VA-conducted urine tests were negative. The New York Daily News reportedly paid for the tests to be repeated, using a more sophisticated technology developed in Germany and the UK. The tests on four of the men were said to be positive for depleted uranium. The soldiers say they were never told of the dangers of DU, that the Department of Defense (DOD) claimed it is safe.[20] The DOD reportedly has another 1.5 billion pounds of DU stored in hazardous waste sites across our country.[21]

The fifteen-nation European Union has taken a completely different stance from the US in its regulation of toxic, and potentially toxic, substances; preferring to err on the side of caution. In the EU, consumer and environmental protection trumps corporate self-interest.[22] Recent EU regulations called REACH, for registration, evaluation, authorization and restriction of chemicals; requires registration of 30,000 chemicals in use in industry. Most EU businesses think that the costs of regulation are a price worth paying. Health benefits of REACH in 2003 were estimated at $50 billion euros over thirty years, vastly overriding estimated costs of three to five billion euros over 11 years.[23] If universal health-care existed in the US, wouldn't we be more likely to consider the health costs of toxic exposures; and conversely, the health benefits of effective regulation of toxic substances?

By contrast, the US Environmental Protection Agency (EPA) takes a back seat, preferring voluntary industry self-regulation. In 2001, EPA apparently recognized that azinphos-methyl, AZM, posed "unacceptable" risks to farm workers, but it postponed stopping its use for four years. Five years later, EPA again appears to recognize "unreasonable adverse effects" of this highly neurotoxic organophosphate insecticide determined to be dangerous to the human brain and central nervous system, respiratory system, and cardiovascular system; as well as dangerous to the environment and ruled that it must be banned, but then continued to allow its use for another six years.[24] EPA explained its actions thusly: "phasing out AZM will encourage growers to use other safer pesticides."[25] EPA's obligation to protect human health and the environment was overridden by corporate interests.

At the same time, EPA has developed a program with Head Start to "educate families on the need to protect children from secondhand smoke ... (and) encourage parents, teachers ... to Take the Smoke-free Home Pledge."[26] All very nice, but again, a smoke-free and toxin-free environment for children could better be achieved by stopping industrial toxins that damage children's brains and central nervous systems, and in some cases, end their lives.

As long as OSHA and EPA are not working together, not sharing full information, not focusing on protecting workers and the environment, respectively; as long as businesses are not paying for the damage to health, life, and the environment that they are causing; this rampant, wanton destruction of life and health in America will continue.

"Rebalancing the power between workers and investors in North America would also require an extension of the rights of citizens to minimum levels of health, safety and conditions of work, ... food free from contamination and minimum levels of clean air and water as well as other fundamental environmental conditions."[27]

"The proper guardians of the public interest are governments which are accountable to all citizens."[28]

References

1. Gupta, Parul. 21 years later, no solace. Agence France-Presse, cited in South Florida Sun-Sentinel, 12/3/05, p. Al.

2. Barab, Jordan. Libby Asbestos Activist Dies; Residents Consider Buyout. January 23, 2007 Confined Space http://spewingforth.blogspot.com/2007/01/libby-asbestos-activist-dies-residents.html.

3. Executive Summary, IG Report, Libby Asbestos, EPA Region 8. U.S. Environmental Protection Agency, undated, www.epa.gov/region8/superfund/libby/igexsum.html.

4. DePalma, Anthony. Love Canal cleanup complete. New York Times, cited in South Florida Sun-Sentinel, 3/18/04, p.3A.

5. Dye factory, cancer deaths a bitter legacy. Loveland Colorado Reporter-Herald, 5/13/06, p.B5.

6. Press Release. Board of Selectmen, Town of Ashland, Massachusetts. May 3, 2006.

7. Ashland Nyanza Health Study. Executive Summary. Massachusetts Department of Public Health, Center for Environmental Health, April 2006, www.massgov/Eeohhs2/docs/dph/environmental/investigations/ashland_final_report06_summary.doc.

8. U.S. Environmental Protection Agency, Region 7, Fact Sheet, Proposal to Find Missouri Lead Plan Substantially Inadequate, Doe Run Company, Herculaneum, Missouri. December, 2005; and State of Missouri v. Doe Run Resources Company, Case No. CV301–0052C-J1, in Circuit Court, Jefferson County, Missouri, July 12, 2002; and Hiles, Sara Shipley; Guevara, Marina Walker. Lead Astray: What happens when an American company offshores pollution? Mother Jones Magazine, November/December 2006, p. 58–60.

9. Williams, J. Lead-laced soil: Southern Missouri and Doe Run face Heavy Issues Together. Soils and the Environment, Vol.6,No.1, University of Missouri School of Natural Resources, 2005.

10. Fox, Maggie. Unborn Babies Soaked in Chemicals, Survey Finds. Reuters, 7/14/05 cited on www.truthout.org.

11. American Brain Tumor Association. Pediatric Statistics. www.abta.org/primer4.htm.

12. Carrero Galarza, Milton. Shared Pain. South Florida Sun-Sentinel, 11/23/03, p. E1.

13. Dudley, D. Margie E. Richard: Pollution Fighter. AARP Magazine, January/February 2006, p.51.

14. General Environmental Exposures and Cancer. National Cancer Institute, www.dceg.cancer.gov/occu-general.html.

15. Pollan, Michael. The Vegetable-Industrial Complex. New York Times Magazine, 10/15/06, p. 17–20.

16. Food Safety Office, Centers for Disease Control and Prevention, www.cdc.gov/foodsafety/

17. Holes in the food safety net. The Washington Post, cited in Daily Hampshire Gazette, 12/12/06, p. Al.

18. Antibacterial compounds taint farms. Los Angeles Times, cited in Loveland Reporter-Herald, 5/13/06.

19. Information from the Statement of Asaf Durakovic, MD, Ph.D., on the Medical Implications of Depleted Uranium. Dr. Durakovic has been a Professor of Nuclear Medicine and Radiology, with thirty-five years research and clinical experience on radioisotopes. www.interactorg.com.

20. Gonzalez, Juan. Poisoned? Shocking report reveals local troops may be victims of America's high-tech weapons. The New York Daily News, April 3, 2004, p.l.

21. Are radioactive weapons sickening U.S. troops? Associated Press, New York in Daily Hampshire Gazette, August 12,13, 2006, p. B7.

22. Mitchener, Brandon. Increasingly, Rules of Global Economy Are Set in Brussels. Wall Street Journal, 4/23/02, p.Al.

23. Regulatory over-reach? The Economist, 12/9/06, p. 70.

24. NIOSH Pocket Guide to Chemical Hazards. Azinphos-methyl, www.cdc/gov/niosh/npg/npgd0044.html; and Say "No" to Six More Years of Deadly Pesticide. United Farm Workers, February 6, 2007, www.ufwaction.org.

25. Azinphos-Methyl Phaseout. US EPA, www.epa.gov, 11/16/06.

26. Smoke-free, Asthma-Friendly Homes for Head Start Families, www.epa.gov/iag/headstart.

27. Faux, Jeff. The *Global Class War*. John Wiley & Sons, Inc., Hoboken, New Jersey. 2006, p.241.

28. A survey of corporate social responsibility: The ethics of business. The Economist, 1/22/05, p. 22.

CHAPTER 12

▼

WHERE DO WE GO FROM HERE?

The time is long overdue to re-evaluate a system that evolved over one hundred years ago; and clearly doesn't meet the needs of injured, ill, or toxic chemical-exposed workers, or the families of workers who died from their work.

Recommendations:

Recommendation 1. Require all governmental programs to move swiftly and appropriately to serve the occupational injury/illness victim, but secure reimbursement from the workers' compensation insurer and/or the employer.

A conservative estimate of the costs of worker injury, illness, and fatality, both direct and indirect, is between $183 billion and $275 billion per year.[1] On average, only 27% of these costs are borne by workers' compensation.[2]

As discussed in the Introduction, the preponderance of all workers' compensation costs, 44%, are borne by injured workers and their families.[3]

Another 10% represent costs that are shifted onto other health insurance.[4]

Eighteen percent, on average, of worker injury, illness and fatality costs are paid by taxpayers; through governmental programs, such as Social Security, Disability, Medicaid, Medicare, and public assistance.[5] The application process for Social Security Disability has been made difficult, lengthy, and costly for the worker disabled by occupational injury or occupational toxic exposure, who often has to apply five or six times before acceptance; while employers are able to walk away from the damage they have caused.

By allowing corporations and their workers' compensation insurers to push their costs onto government programs, our government is depleting funds in Social Security and Medicare, while providing welfare to the corporations. Instead, Medicaid, Medicare, Social Security, and other government programs should move swiftly and completely to serve the injured, ill worker; but require compensation from the employer's workers' compensation insurer for the costs of every occupational injury or illness for which they are responsible, with additional penalties for delays. Hold the corporations and insurers accountable.

If workers' compensation insurers are forced to pay; they, in turn, will require the employer to pay through higher premiums for the reimbursed government dollar outlay for occupational injuries, toxic exposures, fatalities the employer has caused, and which the insurer will have had to pay. This puts far more clout behind the process than any worker could possibly muster; and helps to quell the stalling tactics that benefit the insurer, while throwing the injured worker and family into destitution.

Recommendation 2. Eliminate the separate system of healthcare for occupational injuries, toxic exposures, and disease.

One must ask why a system of medical treatment is based on the payer's needs, rather than the diagnostic and treatment needs of the patient. The major purpose of this duplicative, and duplicitous, medical system has been to save money for the employer and the insurer.

What has been created in workers' compensation in most states is a separate system of healthcare, one that defies the most basic tenet of the medical profession to do no harm.[6] When the only physicians who see work-related injuries and illnesses are the company physicians, and/or the physicians beholden to the insurance carriers; it mitigates against the provision of adequate, appropriate, timely medical care. The employer and insurer are controlling the medical diagnosis and treatment, restricting treatment to the cursorily palliative, or delaying and denying treatment altogether, with impunity. At the same time, a worker's health insurance will refuse to pay when it is a workers' compensation issue.

Thus the injured, ill worker routinely is left without proper medical treatment, without recourse; while at risk for more severe sequelae as a result, and depleting family savings by being "starved out."

But what if healthcare were provided by healthcare institutions and practitioners of the worker's choosing; and all health insurance, all health care institutions were required to provide the necessary and appropriate medical care, and then secure payment from the workers' compensation insurer and/or the employer? Then the injured, or toxic chemical-exposed, worker would be able to receive appropriate diagnosis and treatment in a timely manner.

As reported in Chapter 8, in California in 1992, the State's Insurance Commissioner, John Garamendi, proposed integrating the medical component of both automobile and workers' compensation insurance into traditional group health insurance, for cost efficiencies and reduction of duplication of service. The injured worker would have been able to choose their own treatment provider and would be assured that treatment costs for the occupational injury/illness would have been covered. The projected savings would then help finance "universal health care coverage while improving disability benefits and rationalizing the entire system." The proposal never made it through the legislature. The workers' compensation insurance lobby was too powerful.[7]

Our current health care system costs about 15% of GDP. "Other rich nations devote just 9 or 10 percent of national income to health care, while insuring everyone and enjoying longer life spans and lower infant mortality rates."[8] Administrative costs represent 31% of healthcare costs in the United States.[9] Medicare's administrative cost is only four percent.[10] An independent advisory panel established by Congress concluded that the government pays an additional 11% per patient in Medicare managed care, over the comparable patient in traditional Medicare.[11]

A national healthcare system also creates the incentives for prevention, as well as for proper care for the injured.

Recommendation 3. OSHA: A system in need of overhaul.

The Occupational Safety and Health Administration (OSHA) is the governmental body with responsibility for ensuring the safety and health of most of the 138 million workers in America. OSHA is woefully inadequate to the task and desperately needs to be changed. But instead of strengthening OSHA, the Bush administration has made a push for voluntary compliance over enforcement.

Discussed elsewhere throughout this book is the serious impediment to getting a safety violation declared "willful" and making it stick. The New York

Times examined every workplace death case deemed willful, 1798 of them, both federal and state OSHA cases, over a 20 year period. Only 196 were referred to prosecutors. Only 81 cases resulted in conviction and a total of 16 jail sentences. The maximum sentence for a willful violation of a specific safety standard under OSHA that causes death of a worker is six months in jail. Worse, it's a mere misdemeanor. Harassing a wild burro on federal lands can result in a one-year sentence.[12]

Some states have increased penalties beyond the six months, and three states have made it a crime to commit violations that result in severe injury, as well as death.[13] In California, conviction can result in a three-year sentence and fines up to $1.5 million. This has resulted in a substantial increase in prosecutions over other states and ten times the federal prosecution rate, while substantially lowering the worker death rate, despite lingering resistance to pursuing prosecution in many of the rural California counties.[14] Other states, Florida is an example, have gone in the opposite direction, making it a more difficult fight for injured/ill workers and the families of deceased workers to be compensated.

OSHA fines for a "high gravity" violation, defined as "situations involving danger of death, or extremely serious injury, or illness", are set at $5000. or greater. Up to a 60% reduction is permitted for "size of business," and a 25% reduction for "good faith."[15] In 2003, fines for willful violations averaged $31,763.[16] Compare that with the $331 million stated to be paid by MCI-World Com to settle accounting fraud accusations.[17]

Over the ten-year period ending in 1999, OSHA reportedly deleted almost 14,000 violations per year, thus also reducing the number of repeat and willful violations.[18]

OSHA's budget in FY 2003 was $251.9 million, with 40–50% of dollars earmarked for enforcement.[19] In 2004, OSHA had 2236 inspectors to handle the more than 6 million workplaces within its purview, two hundred less than the agency had in 1975.[20] To inspect each workplace under its jurisdiction just once would take federal OSHA 106 years. In some states, it could take more than 150 years to inspect all job sites just once.[21] There are more federal wildlife protectors than OSHA inspectors—a sad commentary on how we value worker health and life.[22]

Over eight million state and local government employees, including police and fire fighters, are not even included in the safety and health regulatory coverage provided under OSHA. Those who work in air transportation and agriculture industries also lack full protection under OSHA, presumably covered by other laws.[23]

In 2001, the Bush administration revoked an OSHA regulation that barred companies with repeated workplace and environmental health and safety violations from receiving government contracts. In one year, the GAO reportedly logged $38 billion in federal contracts given to 26 contractors with 5,121 violations, and an additional eighty firms that violated labor laws received $23 billion in taxpayer funded projects.[24] What a message our government is sending about the importance of worker safety and health!

In an attempt to appease the auto industry, at the expense of workers health and lives, OSHA took six years to issue a safety bulletin about "lethal levels of asbestos" in brakes; long after the OSHA scientist who researched the matter had completed his work. After finally being pressured by Senator Patty Murray (D-Wash) and the OMB to release the safety information, OSHA' s Administrator, John Henshaw, had the agency also include auto industry-financed studies that determined that asbestos in brakes posed little risk. Sadly, we remain one of the few advanced nations that still hasn't banned the import and use of asbestos products.[25] About 215,000 workers are exposed to asbestos in this country.[26]

Every regulation that OSHA has attempted since 2001 has been withdrawn, including PELS (permissible exposure limits) on air contaminants, and process safety management of highly hazardous chemicals.[27] Since its inception in 1971, OSHA has established permissible exposure limits (PELS) for approximately a mere 300 chemicals out of the 650,000 hazardous chemicals to which an estimated 32 million workers in American workplaces are exposed, and most of the PELS it has established are dangerously outdated.[28]

Limited budget and resources, difficult political battles for tougher regulatory standards, ridiculously low fines, low probability of inspection, miniscule possibility of prosecution and conviction, negotiated agreements for reductions of "willful" to "unclassified" violations, and agreements to no admission of wrongdoing; all this adds up to no incentive for employers to provide a safe and healthy workplace in compliance with standards, no deterrent to flagrant violations. It is tantamount to a license to kill workers with impunity. As we have noted elsewhere, neither workers, nor families of deceased workers, have access to OSHA in either its negotiation or its decision making process, while OSHA seeks out the advice and cooperation of the employers that it is mandated to regulate. OSHA has become the toady of the corporations.

Many times, OSHA's response to FOIA requests (requests under the Freedom of Information Act) excludes fatality reports, employee interviews, witness statements, correspondence and memoranda from the case file. Thus families, media, the public are kept in the dark.

Additionally, OSHA and BLS, both agencies within US Department of Labor, collect data on companies and their injury and illness rates. While both agencies are collecting overlapping information, neither has the full picture, since millions of employers are excluded from one or both data sources.[29] BLS won't release specific company information to OSHA, out of fear that employers won't report accurate data. OSHA thus began its own Data Collection Initiative (DCI); which it uses to target work sites with high rates of injuries, or illnesses, or both. OSHA does not release this information to the public. It would cost nothing, and would benefit the health of workers and communities throughout our country, to release this information to the public.

Cullen (2002) points out that this "lack of public oversight has contributed to the deplorable condition of our nation's workplace health and safety. Turned around, it can also lead to the cure."[30]

OSHA must be allowed to raise the fines at least on a par with those for accounting fraud and other corporate crimes. Raising fines would generate substantial revenue, which the agency could use to do its job. OSHA needs to be budgeted and staffed to enforce its safety and health mandate.

The process for approving and upgrading PELS, or improving standards, must be streamlined and overhauled. It is simply untenable that hundreds of thousands of hazardous chemicals are allowed in use in the workplace, yet are unmonitored, and unrestricted. Saving the lives and health of American workers needs to be a governmental priority.

Unions, workers, family members, environmentalists, epidemiologists, toxicologists, public health, occupational medicine physicians, and researchers, all must be involved in OSHA's rule-making and oversight process. As Tammy Miser, founder of United Support and Memorial for Workplace Fatalities, states: "We are the brothers, sisters, mothers, fathers and children of America. We are not a number or a statistic, we are the very ones making your profits. We are not asking for more than we are entitled to; our right, the right to a safe and healthy workplace. We need leaders to restore our faith in governmental humanity." Miser and others have drafted a *Workplace Tragedy Family Bill of Rights*.

The time to overhaul OSHA is long overdue: overhaul not to the benefit of the corporations, which is what is being done now; but to perform the tasks for which it was founded, to assure the safety and health of the workplace for workers, to refer cases of depraved indifference to worker safety and health to Department of Justice for prosecution.

Recommendation 4. Establish a national medical and statistical data base on worker injuries, worker toxic exposures, and occupational diseases.

Incredible as it sounds, there is still no national comprehensive surveillance system for occupational injuries, occupational diseases, nor for the diseases and deaths whose etiology was occupational toxic exposure. The partial and fragmentary systems that do exist grossly underestimate incidence, are unreliable, and inconsistent.[31]

Leigh, et al.(2000) had to research hundreds of primary and secondary data sources in order to arrive at a conservative estimate of occupational diseases and the deaths that they caused, because there was no single governmental source for this data.[32]

As noted in the Introduction, estimated deaths from occupational diseases are more than deaths from breast, or prostate, or colon and rectal cancers alone, or from Alzheimer's Disease, or heart failure, or HIV. The costs of occupational injuries and illnesses are also higher, commensurate with their numbers and severity. But because the magnitude of occupational injuries and diseases is not known, "the national debate on medical care rarely addresses occupational safety and health issues."[33]

As we lag behind in a national database, we have also lagged behind in research, often attributing cancer causation to smoking, but not to toxins in the workplace. An estimated 60,000 lives are lost each year to worker toxic exposures. In the fatality data, it all gets subsumed in the International Classification of Diseases (ICD) category labeling the final cause of death, with no attention to the toxic exposure that caused the body breakdown, that resulted in death.

Data gathering, aggregation, and surveillance are essential to assist in diagnosis, treatment, improved prognosis and prevention; and are severely hindered by the lack of a single, comprehensive, national surveillance system. As we have seen, the provision of appropriate diagnosis and treatment suffers from the lack of medical knowledge about toxic exposures, occupational illnesses, and attendant disease processes.

For the first responders working in "the pile" at the World Trade Center, it reportedly took five years for clinical guidelines to be issued for the diagnosis of their 9/11 related illnesses, thus hampering diagnosis and treatment.[34] Thirty percent of these workers appear to have been misdiagnosed, e.g., being told they had asthma, when their lungs contained pulverized glass.[35] Firefighters and emergency personnel have experienced lung function decreases equal to twelve years of life, from their brief time working at ground zero. The future is equally grim: diseases such as cancers and asbestosis may not show up until decades after exposure.[36] While the individual toxins in the ground zero stew were already known, exposures to this particular constellation of toxins was unknown prior to 9/11, also hampering the ability to diagnose and treat.

Under NIOSH, Education and Research Centers (ERCs) are funded to provide institutional support and training programs in occupational medicine, epidemiology and toxicology. Under-funded and narrowly technical, the potential is there "to develop and nurture a new core of well trained, knowledgeable, articulate, and committed scientists who accept the challenge of building bridges between public health science and public health policy."[37]

A national comprehensive occupational disease surveillance database would lay the groundwork for health care and academic institutions' continued research into the causes and consequences of occupational illnesses and injuries.[38]

Recommendation 5. The Seventh Amendment—for workers too.

The Seventh Amendment doesn't exist here. Yes, it's been a part of the Bill of Rights, which we so proudly hail, since 1791.[39] The Seventh Amendment to the Constitution guarantees the right to trial by jury to everyone in America—everyone, that is, except workers injured, made ill or killed by their work.[40]

How did this happen? As discussed in the Introduction, the quid pro quo is supposed to be swift and certain wage compensation and treatment in exchange for having given up the right to sue. As we have seen, the immunity from tort liability has evolved to a protection of insurers and employers; and a depraved indifference to the health and safety of workers.

Workers would stand a far better chance of being in a safe working environment if they could sue, just like any other injured party. Workers' compensation should be one alternative; which could be accessed, as long as the worker or the family of the deceased worker, are satisfied that they are being treated fairly. Nor should it be short-circuited by acceptance of beginning treatment or wage compensation. To do so would be an invitation for employers and insurers to continue to pursue delay and deny tactics. Justice delayed is justice denied, and workers' compensation as the exclusive remedy is surely not justice. The constitutional right to sue must be inviolate. To continue with workers' compensation as the exclusive remedy is to protect employers and insurers from accountability for the disease that cripples, or the corporate negligence that kills.

Following a refinery explosion in California, the families of three workers who died in the explosion reportedly were awarded $21 million. These workers had been employed by a subcontractor at the site, and thus were able to sue the refinery for negligence. But an employee, stated to have barely survived after jumping sixty feet from a blazing tower, and breaking almost every bone in his body found that his sole remedy was workers' compensation. He is reported to have endured 24 surgeries, amputation of thumb and fingers on one hand, and numerous skin grafts, and be confined to a wheelchair, with metal pins in his knee and thigh. By contrast, the employee appears to have been awarded a total temporary disability benefit of $490. per week.[41]

As this case demonstrates, workers' compensation is not an equivalent remedy. How can it be, when injured workers and their families are often left destitute; while the insurers of those who caused the injuries, the toxic exposures, the deaths pay only 27%, on average, of the costs of those injuries, toxic exposures, deaths? How can it be, when everyone else has the right to sue, but not the injured worker?

As Cullen points out, the workers' compensation exclusive remedy system works only for the minor injury or brief illness.[42]

Recommendation 6. Criminal liability for health and safety violations.

The Sarbanes-Oxley Act was quickly passed following the Enron scandal. By 2003, fraud investigations into numerous corporations and substantive criminal charges against their executives, employees, and owners were in process.[43] It is reported that Walter Forbes, of Cendant Corporation, received a twelve-year prison sentence and $3.275 billion restitution order; Jeffrey Skilling, of Enron, a 24 year and four month sentence; and Ebbers, of WorldCom, a 25 year sentence, all for accounting fraud. [44]

Sarbanes-Oxley increased criminal penalties for crimes of fraud and cover-up, and strengthened whistleblower protections.[45] Penalties to individuals found guilty of conspiracy to commit financial reporting fraud were increased to a maximum ten years in prison, and a maximum fine of $1 million. A willful violation could result in a twenty-year prison term and fines up to $5 million.[46] Compare this to the paltry OSHA fines, and rarely imposed misdemeanor jail time, for the repeated violations that kill and maim workers.

Canada and Australia have enacted Corporate Manslaughter Laws, and the UK has drafted similar legislation. [47] One way to proceed in America would be to expand the provisions of Sarbanes-Oxley to include criminal liability for health and safety violations in the workplace.

We can learn, both from the weaknesses that have been identified in the draft Corporate Manslaughter Bill for England and Wales, as well as the far more effective principles of liability incorporated into the reformed criminal code in Canadian law.

A few of the flaws that have been identified with the UK draft, are as follows:

Only deaths that occur in the workplace would be covered; not serious injuries, nor deaths from occupational diseases.

The maximum penalty would be an unlimited fine, no prison time.

The larger the company, the more difficult the prosecution; due to the requirement that first, senior official(s) be identified as responsible for the management failure, as well as proof that senior management would have profited from the proven gross negligence. Instead, corporate liability should be based on management failure; and the evidential burden then falls on the organization to prove that it had acted with due diligence to prevent the corporate culture that "encourages, tolerates or leads" to worker death.[48]

In Canada, An Act to Amend the Criminal Code, C-45, became effective in 2004. In C-45, Canada chose to apply a new principle to the criminal offenses already extant in its Criminal Code, thereby making it possible to attribute criminal liability to corporations and other organizations.

The Canadian law makes it possible to convict the organization on the basis of the actions (or failure to act) of any employee and a "relevant senior officer;" the definition of the latter to include "members of management with operational, as well as policy-making, authority."

"For crimes of intent or recklessness, criminal intent will be attributable to a corporation … where a senior officer is a party to the offence, or where a senior officer has knowledge of the commission of the offence by other members of the organization and fails to take all reasonable steps to prevent or stop the commission of the offence."

For crimes of criminal negligence causing bodily harm or death, the organization will be guilty where the senior officer responsible, or senior officers of the organization in the aggregate, "show a marked departure from the reasonably expected standard of care in failing to prevent a representative (of the organization) from being a party to the offence."

C-45 adds a new Section 217.1 to the Canadian criminal code to facilitate the "application of the offence of criminal negligence," by clarifying the existence of a legal duty on the corporation and senior officers: "those who are responsible for directing the work of others are under a legal duty to take reasonable steps to prevent bodily harm to any person arising from such work."

Section 732.1 (3.1) allows probation orders to be served upon corporations following conviction, with additional conditions, such as making restitution to the victims for any loss or damage suffered resulting from the offense, remedial measures to prevent any further safety or health offenses, communicating the new policies and procedures to employees, reporting to the court and appropriate provincial oversight departments; and requiring the corporation to inform the public of "the offence of which the organization was convicted, the sentence imposed, any measures taken by the organization to reduce the likelihood of its committing further offences." There is no limit on the fines that can be imposed for indictable offenses.[49]

Two components that should be included in an American Corporate Manslaughter Law would be equity fines, i.e., company shares that would go into a compensation fund for injured, ill, or deceased workers; and "adverse publicity orders," the latter requiring a convicted corporation to advertise their conviction, affecting their corporate reputation.[50]

One argument that has been raised against implementing such a law has been that directors and companies will leave the country if they are to be held so accountable for their negligent behavior. Neither Canada nor Australia appears to have experienced any such flight since their laws were passed.[51] "Criminal sanctions can and must play an important role in deterring irresponsible business decisions;" and in fact have been shown to do so, because business crime is rational rather than impulsive.[52]

Surely, if Congress can act to find broadcasters culpable for indecency and impose fines up to $325,000 per incident,[53] it can act to hold corporations accountable for corporate manslaughter. If we can hold corporations accountable for accounting fraud, we can hold them accountable for criminal negligence that destroys the health and life of workers. Make companies liable for the depraved indifference that kills.

Finally, as recommended by National Council for Occupational Safety and Health (National COSH), "local district attorneys should prosecute employers whose actions result in worker deaths to the fullest extent possible under state and local criminal law."[54]

Recommendation 7. Saving the environment demands rigorous regulation of industrial toxins.

See Chapter 11.

Recommendation 8. Setting fair wage compensation under Workers' Compensation.

Wage compensation under Workers' Compensation is set by states; which have increasingly been responsive to the pressure of workers' compensation insurance and business lobbies to reduce the rate, until today wage compensation to injured/ill workers in sixteen states is below the poverty level. The average weekly wage compensation rate in most states stands at a mere 120% of poverty, with the highest payout of any state at 170% of poverty.[55] At the same time, corporate executive compensation has reached a ratio of 431 to one, or 431 times the average worker's pay.[56]

In California, a study by the Rand Institute for Civil Justice found that at the largest California firms, workers with work-caused permanent partial disabilities received less than half their wages in the five years following their injury.[57]

Within one hundred days of September 11, 2001, the US Department of Justice (DOJ) announced its compensation fund regulations "to provide victims with an alternative to the long and costly litigation process." The fund appears to have been created "to protect the airlines and US entities from possibly crippling litigation...."[58] Advance benefits of $25,000 for those seriously injured and $50,000 for the families of deceased victims were arranged. The minimum compensation for single deceased victims was set at $300,000, for married deceased at $500,000.[59]

By March 7, 2002, the DOJ's final rules were published. The plan attempted to set a swift, non-adversarial, and standardized process based on a formula that took into account the age, annual income, marital status, and number of dependents of each deceased and/or injured person. An additional compensation for pain and suffering was doubled for dependents in the final rules, up to a maximum of $550,000. Originally, all payouts from other sources, e.g., insurance, were to be deducted from the final award. Final rules excluded pension plan payments made directly by the deceased.[60]

By contrast, as discussed in Recommendation No. 4, the rescue workers at the World Trade Center site in New York, who inhaled the toxic chemical stew at Ground Zero suffer from severe respiratory illnesses and face an uncertain future, many without medical coverage.[61]

By 2004, the average death payment from the 9/11 Compensation Fund was reported to be slightly more than $2 million, the range up to $7.1 million. Congress had authorized $5.9 billion.[62] What Kenneth Feinberg, the appointed Special Master and architect of the plan had achieved was to create a methodology for determining compensation for loss of life. No amount of money can compensate the bereaved families for the loss of their family member. But that fact is also true for the estimated 66,000 persons who lose their lives from work injury or occupational toxic chemical exposure each year in America. Compare these compensations to the $5,000 received by the family of Dale Goldstein, or the zero dollars to the family of Shawn Boone because he was unmarried, with no children.

Recommendation 9. Coalition of organizations.

In the past, unions were the major defenders of rights for workers. Now their ranks have dwindled. It will take a coalition of unions and the committed to effectively press for change—a coalition whose voices will be strong enough to rise above the almighty politics of corporate profit.

As Michael Silverstein, MD, MPH states, "safe jobs must become a matter of broad public interest. Workplace safety has been marginalized as an issue of public policy.... [T]his public blind spot must be consciously and systematically addressed."[63]

State-based organizations are essential, since workers' compensation is regulated by states. Often state-wide organizations, that regularly do battle on behalf of workers, find themselves consumed with fighting the latest proposal for legislative or regulatory change brought by the powerful insurance industry, backed by business lobbies; to implement yet another reduction in the already meager benefits, or another hoop to jump through for injured, or toxic chemical-exposed, workers. In that environment, it is difficult to effect positive change for workers.

When janitors in the Service Employees International Union (SEIU) in Houston found themselves in a contentious battle with the cleaning companies over hourly wages and health care coverage; the union reportedly not only pressured the cleaning companies, but also the building owners. The strategy was effective and they won a contract with a substantive raise in wages and, for the first time, the provision of health care coverage.[64]

The Coalition of Immokalee Workers (CIW), an organization representing the tomato workers in and around the community of Immokalee in central Florida, tried to get a one-cent per pound raise from the growers for years, without success. Then they crafted a strategy for developing "autonomous allies:" cadres of social activist groups; including labor, church groups, students, immigrant rights groups. There were hunger strikes and long marches. Over time, the focus became a national boycott of Taco Bell, a major purchaser of tomatoes. As part of the boycott, a campaign led to the removal of Taco Bell from many high school and university campuses, following pressure from student groups. After a four-year-long boycott, Taco Bell's parent company, Yum! Brands Inc., reportedly agreed to the increase, to go directly into farm workers' wages; and a commitment by Taco Bell to "seek new laws that better protect all Florida tomato farm workers." It's a true David and Goliath story: how this small local organization of tomato farm workers; living in poverty, in the heart of the plantation-like growing country of central Florida; appears to have succeeded in getting the world's largest restaurant company, Yum! Brands, to meet every one of its demands, in what is described as a first-time-ever agreement. Workers know that securing safe conditions and decent wages really requires that workers be involved in the decision-making process. Everything that CIW does is group-centered, with "strategies fully developed at the base." It must also be noted, that the use of this highly successful secondary boycott, i.e., a boycott against an entity other than the employer, would have been illegal to use had CIW been a union.[65]

Jobs With Justice (JWJ) was formed in 1987, to incorporate the struggle for workers' rights into a larger campaign for social and economic justice. Now throughout the country, JWJ is a coalition of labor, community, student and religious organizations committed to workers' rights; explaining that workers are also "neighbors, members of communities, people of faith, students, and parents of students, taxpayers, caregivers, heads-of-households, and family members. Assaults on workers' rights are also assaults on the stability and well being of neighborhoods, religious communities, school systems and universities, and families."[66] The JWJ has established Workers' Rights Boards and a National Workers' Rights Board, comprised of shakers and movers: legislators, academics, religious leaders, activists "to provide a collective and unified voice in support of workers," and to set community standards for behavior of employers.[67]

Recently scientists and evangelicals have found themselves on the same side on global warming. One minister reportedly put it this way, "Science can be an ally in helping us understand what faith is telling us. We will not allow the Creation to be degraded, destroyed by human folly."[68] If faith-based organizations and scientists can align on global warming, surely they can become a strong voice to stop the environmental devastation caused by toxic chemical exposures in the workplace. As one workers' compensation attorney pointed out, worker safety and health is a family values issue.

Tammy Miser lost a family member in a workplace fatality (see Chapter 5). Out of her grief and determination to make a difference, she and a colleague created United Support and Memorial for Workplace Fatalities (USMWF). Organizations such as USMWF have an important role to play in giving grieving families a place to turn. To paraphrase mine safety activist, Mary Harris Jones, a.k.a. Mother Jones: grieve "for the dead, and fight like hell for the living." USMWF has become an important source of information and collaboration on both national and international worker safety and health issues.

Other websites have also served as resources in covering workplace safety and health issues and the workplace injuries and deaths that are occurring throughout America. The Weekly Toll at http://weeklytoll.blogspot.com, also operated by Tammy Miser, provides weekly news and updates on fatalities in the workplace. The Weekly Toll also links to Congressional bills on worker health and safety issues, and has links to other health and safety websites. The aforementioned Bill of Rights for family-member victims of workplace fatalities, entitled *Workplace Tragedy Family Bill of Rights*, can be found at www.usmwf.org. Confined Space @TPH (The Pump Handle) posts occupational health and safety news and occupational injury, fatality incident information. Other significant sources are Hazards Magazine, a well-respected UK publication with a global view, at www.hazards.org; and LabourStart, an international health and safety news source by country, at www.labourstart.org

Trade unionists around the globe observe April 28th as Workers' Memorial Day, an international day of mourning for the workers who die from workplace injuries and illnesses each year. April 28th was chosen because it is the day of a similar remembrance in Canada, as well as the anniversary of the start of OSHA. Greater publicity could help to bring the need for stronger safety and health standards and enforcement into the public consciousness.[69]

As the economy becomes a global one, the importance of organizing globally becomes paramount. As Jeff Faux (2006) points out, the investor class is organized and protected globally. Workers for the most part are not, and consequently, the environment isn't protected either. This is increasingly true since NAFTA (North American Free Trade Agreement) and the World Trade Organization (WTO), which added trade sanctions to the power of money for global corporations.[70] The WTO actually prohibits any laws that protect the environment, workers or public health if they "interfere with the freedom of corporations to invest, buy and sell." It gives the WTO the power to override any nation's laws, to challenge those laws if it "can be shown to impair the benefits that the corporation could expect to receive under the WTO."[71]

The International Confederation of Free Trade Unions (ICFTU), represents 150 million workers, or one-eighteenth of the world's workers. The term "free" means the confederation is comprised mostly of unions independent of the company or the government. Such a confederation can share information, and provide support and cooperation across nations and continents, as e.g., when rubber workers in Turkey held a half-day work stoppage to support workers in the United States, reportedly on strike against Bridgestone-Firestone. At the same time, imprisonment and state-sponsored violence against unions continues in many countries.[72]

The National Council for Occupational Safety and Health, which has tended to be a consortium of state-based affiliates, increasingly is reaching out and connecting across continents, establishing links with the European Work Hazards Network, the Trade Union Councils, et al. National COSH is a young organization, and this is a good start.

As Jeff Faux(2006)notes, what is needed is to establish protections for workers and the environment, comparable to the protections afforded investors.[73]

Recommendation 10. Insurance regulation.

Workers' compensation insurance is regulated by individual states. State insurance commissioners are generally state governor appointees. Where do they go after their tenure? Back to the insurance industry from whence they came, the insurance industry that they once regulated. Coupled with the heavy pressure applied by the workers' compensation insurance and business interests to reduce worker benefits and weaken worker protections, it all creates what AFL-CIO describes as a "constant race to the bottom" for workers.[74]

Often not factored into a state's premium rate setting is the insurer's stock and bond market investment income, and the hefty underwriting salaries they pay to themselves.[75] State insurance departments tend to rely solely on data supplied to them by the workers' compensation insurance industry they are commissioned to regulate, with no independent audit or review. One state, Massachusetts, mandates independent actuarial reviews of rate filings by insurance companies, and rate setting by the commissioner.[76]

When insurers bring pressure to increase premiums or to reduce worker benefits, nothing is said about the condition of the insurer's investments. What one hears instead is how medical costs are rising, or workers' attorney fees are too high (but not the army of attorneys working for employers or insurers, of course) or that benefits are too generous.[77]

Every state should have an insurance commission or oversight board comprised of workers, family members of injured or deceased workers, union representatives, occupational medicine physicians, attorneys who represent the rights of injured workers, and environmentalists; as well as the employer and insurance representatives that have always had the ear of the insurance commissioner. This oversight commission/board should convene regularly, periodically and have input into all proposed changes in workers' compensation law or regulations being considered.

States should impose heavy and continuous penalties for workers' compensation insurers that delay and inappropriately deny benefits. Delays and denials allow the insurers to keep their money invested. Without serious penalties, there's no disincentive to delay and deny, starving out the injured/ill worker and his/her family, ultimately leading to a settlement out of desperation; ordaining the family to poverty, but saving money for the insurer.

The asbestos industry reportedly funded studies that showed the "causal relationship between asbestos exposure and cancer" as early as the 1940s. Yet this information was withheld from workers, some already showing evidence of irreversible disease.[78]

No workers' compensation settlement should be allowed to be secret. By stifling the employer's dirty little safety and health secrets, it has allowed history to continuously repeat itself, damaging the lives and health of many others. It isolates the injured or toxic-exposed worker victim, so that others cannot be forewarned. In the case of toxic exposures, it puts a veil of secrecy between the toxic-exposed workers and the community, which is also being affected.[79]

Recommendation 11. General occupational injury and illness tax.

As previously noted, workers' compensation covers only about 27% of worker injury, illness costs. Leigh, et al.(2000) recommends "a general occupational injury and illness tax to be levied on all employers" in order to pay for the workers' compensation costs now being borne by families, taxpayers, and the general public. Leigh, et. al.(2000) recommends that the tax be patterned on the Federal Black Lung Trust Fund. That fund taxes the coal companies on a per tonnage basis in order to pay the medical costs of pneumoconiosis, a coal mining-caused disease. Taxes could be assessed based on each industry's contribution to cancers and other specific diseases.[80]

Recommendation 12. Re-thinking how work is done.

In tandem with OSHA's toothlessness, the way that work is being done has changed. Increasingly, production lines are being speeded up resulting in crippling, sometimes deadly, injuries; all in service to profits at the expense of workers' health and lives.

Meatpackers have the highest serious injury rate of any occupation, about 40,000 workers, 27% of its workforce, injured every year. Lacerations, amputations, eye injuries, burns, toxic exposures, or being crushed to death are some of the horrendous injuries and fatalities that result from this industry's dangerous disregard for worker safety and health. The typical production line speed in the industry has more than doubled in the last 25 years; in an industry where one is wielding knives, saws or other power tools, while standing closely together on a wet slippery floor and feeling rushed by a conveyor belt that never stops. It's been reported that even when a worker is cut, bleeding, and collapsed unconscious, the conveyor chain still keeps going.[81]

In many industries, temporary workers are employed to perform what were once full time permanent positions. A corporation thus looks better on paper because worker injuries are attributed to the temporary agency.

Union membership has declined, and along with it, the vigilance on worker safety and health that has existed in union shops. In the non-union environment, workers are unlikely to speak out when they identify a safety or health hazard, for fear of employer retaliation or firing.

We need to rethink the way work is done: worker safety and health must become primary. There can be no justification for allowing the destruction of health and lives in service to profit.

Recommendation 13. Media honesty and thorough analysis.

"the work day was cut short for over forty farm workers after they were exposed to a nontoxic fungicide." (Italics added). This description makes it sound like a bonus to the workers, rather than the exposure to a sulfur-based chemical fumigant requiring twelve hours in hospital, decontamination, and unknown health impact for their lives in the future.[82]

A front-page article in the Daily Hampshire Gazette, entitled *"Worker error led to collapse,"* notes injuries to two workers after they fell twenty feet onto concrete, immersed in a dozen half-ton trusses that collapsed. Both the title of the page one article and the first paragraph put the onus of responsibility for the collapse on the workers who were deemed to be "not properly bracing those structures." If one continues to read, the writer does quote the city's building commissioner that, "the subcontractor ... failed to follow the manufacturer's instructions for the installation of the trusses, which caused the trusses to collapse." But despite this reported acknowledgement of the subcontractor's critical role in the failure, the article attributes the collapse and the injuries to the workers.[83]

Too often, the image one gets from what little media attention is paid to worker injuries, worker deaths, worker toxic exposures is that the worker caused their own injury, or death, or toxic exposure; or that it's a "freak accident," interpreted as no one is to blame. The sheer magnitude of occupational deaths and injuries, the depraved indifference of corporate disregard for worker safety and health, facilitated by government intransigence or outright concealment, is not even on the radar screen.

Media honesty and in-depth analysis in worker injuries, deaths and toxic exposures is needed. In fairness, there are also many fine examples, such as the New York Times three part series based on an eight-month examination of OSHA and worker deaths. More is needed.[84]

*   *   *   *

"By virtually any measure, occupational injuries and illnesses do not receive the same scientific, government, medical, media, or public attention commanded by AIDS, Alzheimer's disease, arthritis, cancer, or heart disease."[85] Yet the sheer magnitude of occupational injuries and illnesses cries out for attention in all of these arenas, and would improve the health of our nation.

"Long after Congress declared safe and healthful workplaces to be a national priority, more attention is paid and more resources are devoted to fish and wildlife protection than worker safety.... We have the technology, the legal framework and the moral capacity to do significantly better."[86]

References

1. Liberty Mutual quoted in *Death on the Job: The Toll of Neglect.* 13th edition. AFL-CIO, Washington, D.C., April 2004, p. 1.

2. Leigh, J. Paul; Markowitz, Steven; Fahs, Marianne; Landrigan, Philip. *Costs of Occupational Injuries and Illnesses.* University of Michigan Press, Ann Arbor, Michigan, 2000. p.2.

3. ibid., p. 175.

4. ibid., p. 175.

5. ibid., p. 175.

6. The actual statement from The Physician's Oath, Hippocrates, c. 460–400 B.C., is ... "I will use treatment to help the sick according to my ability and judgment, but never with a view to injury and wrongdoing ... to help the

sick, and I will abstain from all intentional wrongdoing and harm," cited in Bartlett's Familiar Quotations, Fifteenth Edition, p.78, Little, Brown and Company, Inc., Boston, 1980.

7. Himmelstein, Jay,MD, MPH; Rest, Kathleen, PhD., MPA. Working on Reform. How Workers' Compensation Medical Care is Affected by Health Care Reform. Public Health Reports. January/February 1996, Vol 111, p.24–25.

8. Miller, Matthew. Health Care: We'd rather shift costs than find savings. Tribune Media, cited in South Florida Sun-Sentinel, 1/17/03.

9. Woolhandler, Steffie; Campbell, Terry; Himmelstein, David. Costs of Health Care Administration in the United States and Canada. New England Journal of Medicine. Vol 349:768–775. August 21, 2003.

10. Freccero, Yvonne. The case for universal health care. Daily Hampshire Gazette. 1/17/06, p. A8.

11. Democrats look to cut payments to insurers. Daily Hampshire Gazette, 11/24/06, p. B6.

12. Barstow, David. When Workers Die: A Culture of Reluctance. New York Times, December 22, 2003, p.l.

13. ibid.

14. Barstow, David. When Workers Die: The California Way. New York Times, December 23, 2003. p.l.

15. OSHA Field Inspection Reference Manual CPL 2.103, Section 8, Chapter IV Post-Inspection Procedures, C.2.f.(2) and C.2.h.(3)i.

16. *Death on the Job: The Toll of Neglect.* 13th edition. AFL-CIO, Washington, D.C., April 2004, p. 43.

17. Levy, Marc. MCI will pay $331 million to states in fraud settlement. South Florida Sun-Sentinel, 10/4/05, p. 2D.

18. Cullen, Lisa. *A Job To Die For.* Common Courage Press, Monroe, ME., 2002, p.165.

19. Weil, David. OSHA: Beyond the Politics. Frontline, January 9,2003 at www.pbs.org/wgbh/pages/frontline/shows/workplace/osha/weil.html.

20. *Death on the Job: The Toll of Neglect.* 13th edition. AFL-CIO, Washington, D.C., April 2004, p. 57.

21. ibid., pp. 7, 116.

22. OSHA: A Dangerous Business. Interview: Charles Jeffress. Frontline, January 9, 2003, at www.pbs.org/wgbh/pages/frontline/shows/workplace/osha/ jeffress.html.

23. *Death on the Job: The Toll of Neglect.* 13th edition. AFL-CIO, Washington, D.C., April 2004, p. 8.

24. AFL-CIO Vigorously Opposes Elimination of Contractor Responsibility Rules. AFL-CIO Press Release, 12/27/01, cited in Mogensen, Vernon, Editor. *Worker Safety Under Siege.* M.E.Sharpe, Inc., Armonk, New York, p.23.

25. Parks, James. Bush OSHA Tries to Scapegoat Scientist for Doing His Job. 11/21/06. AFL-CIO News that works now, http://blog.aflcio.org/200 6/11/ 21/bush-osha-tries-to-scapegoat-scientist-for-doing-his-job/.

26. Brandt-Rauf, Paul; Yongliang, Li. P53 Biomarker and Intervention in Occupational Cancer. 2002–2005. Abstract. National Institute for Occupational Safety and Health, www2a.cdc.gov/nora/selectedPjt.asp?id=53.

27. *Death on the Job: The Toll of Neglect.* 13th edition. AFL-CIO, Washington, D.C., April 2004, pp. 47–56.

28. Cullen, Lisa. *A Job To Die For.* Common Courage Press, Monroe, ME., 2002, p.75.

29. ibid., pp.177, 178.

30. ibid.,p.172.

31. Azaroff, Lenore; Levenstein, Charles; Wegman, David. Occupational Injury and Illness Surveillance: Conceptual Filters Explain Underreporting. American Journal of Public Health, September 2002, Vol 92, No.9, p. 1421–1429.

32. Leigh, J. Paul, Markowitz, Steven, Fahs, Marianne, Landrigan, Philip. *Costs of Occupational Injuries and Illnesses.* University of Michigan Press, Ann Arbor, MI, 2000, pp.3,6.

33. ibid., pp.13, 193.

34. DePalma, Anthony. Officials Slow to Hear Claims of 9/11 Illnesses. New York Times, September 5, 2006, cited at www.truthout.org.

35. CBS News. The Dust At Ground Zero. New York, September 10, 2006, www.cbsnews.com/stories/2006/09/07/60minutes/printable 1982332.shtml.

36. Alaya, Ana. Ground Zero workers ill. Newhouse News Service, in the Sunday Republican, September 10, 2006, p.C6.

37. Silverstein, Michael. Getting Home Safe and Sound? OSHA at Thirty-Five. American Journal of Public Health, in press.

38. Weil, David. OSHA: Beyond the Politics. Frontline. January 9, 2003.

39. Amendment VII to the Constitution reads: "In Suits at common law, where the value in controversy shall exceed twenty dollars, the right to trial by jury shall be preserved, and no fact tried by a jury shall be otherwise re-examined in any Court of the United States, than according to the rules of the common law."

40. Zientz, Mark. The Nine Amendments to the Constitution, www.workerscompensationinsurance.com/articles/9amendme nts.htm.

41. Young, Julius. Workers' Compensation Reform: Why Is It Needed? Boxer & Gerson, Oakland, CA, cited in Cullen, Lisa. Frontline. A Dangerous Business: The Myth of Workers' Compensation Fraud. 2003. www.pbs.org/wgbh/pages/frontline/shows/workplace/etc/fraud.html.

42. Cullen, Lisa. Frontline. A Dangerous Business: The Myth of Workers' Compensation Fraud. 2003 www.pbs.org/wgbh/pages/frontline/shows/workplace/etc/fraud.html.

43. Brickey, Kathleen. From Enron To Worldcom and Beyond: Life and Crime after Sarbanes-Oxley. Washington University Law Quarterly, 81:357–401, 2003, p. 358.

44. Ex-Cendant official gets 12 years, $3.2B fine. Associated Press, Bridgeport, CT in Daily Hampshire Gazette, January 18, 2007, p. B6.

45. Brickey, Kathleen. From Enron To Worldcom and Beyond: Life and Crime after Sarbanes-Oxley. Washington University Law Quarterly, 81:357–401, 2003., p.359.

46. 18 USC 1348, 1349, 1350, cited in Brickey, Kathleen. From Enron To Worldcom and Beyond: Life and Crime after Sarbanes-Oxley. Washington University Law Quarterly, 81:357–401, 2003, p.379.

47. Johnson, Chris. Ten Contentions of Corporate Manslaughter Legislation: Public Policy and the Legal Response to Workplace Accidents. Department of Computing Science, University of Glasgow, Glasgow, G129QQ, Johnson@ dcs.gla.ac.uk www.des.gla.ac.uk/~johnson/papers/Chris_Corporate_ Manslaughter. pdf.

48. Whyte, Dave. Taking offence. Hazards Magazine.PO Box 199, S1 4YL England, www.hazards.org/deadlybusiness/takingoffence.htm.

49. All the information on C-45, Canadian law on the Criminal Liability of Organizations is from Bill C-45: An Act to Amend The Criminal Code, prepared by David Goetz, Law and Government Division, Legislative Summaries, Library of Parliament, Parliamentary Information and Research Service, www.parl.gc.ca/common/ Bills_ls.asp?lang=E&Parl=37&Ses=2&ls=C45&source=Bi, and Bill C-45— Amendments to the Criminal Code Affecting the Criminal Liability of Organizations, Department of Justice, Canada, www.canada.justice.gc.ca/ en/dept/pub/c45/index.html.

50. Bergman, David. Manslaughter bill lets top directors escape justice. No real convictions. Hazards Magazine, P.O. Box 199, Sheffield, SI 4YL England, 2005, www.hazards.org/deadlybusiness/noconviction.htm.

51. Whyte, Dave. Taking offence. Hazards Magazine. PO Box 199, SI 4YL England, www.hazards.org/deadly business/takingoffence.htm.

52. Braithwaite, John; Geis, Gilbert. On Theory and Action for Corporate Crime Control. Crime and Delinquency 28 April 1982, pp. 300–305, cited in Cullen, Francis; Cavender, Gray; Maakestad, William; Benson, Michael. *Corporate Crime Under Attack. The Fight to Criminalize Business violence.* Anderson Publishing, LexisNexis Group, 2006, p. 304.

53. Abrams, Jim. Broadcasters to pay dearly for indecency on programs. Associated Press in South Florida Sun-Sentinel, 6/8/06, p.3A.

54. Campaign to Stop Corporate Killing. National Committees/Coalitions on Occupational Safety and Health, www.coshnetwork. org/national_network.htm.

55. Hunt, A., et al., Adequacy of Earnings Replacement in Workers' Compensation Programs, Washington, D.C., 2004, cited in A National and State-By-State Profile of Worker Safety and Health in the United States. *Death on the Job: The Toll of Neglect.* 13th Edition, AFL-CIO, Washington, D.C., April, 2004, pp.16, 17, 117.

56. Too many turkeys. The Economist, November 26, 2005, p.75.

57. Compensation for Worker Injuries Found Similar at Large and Small Firms. Outcomes Are Inadequate and Inequitable in Both Cases, RAND Analysts Say. RAND News Release, August 30, 2000, www.rand.org.hot/Press/workerscomp.html.

58. Barrett, Devlin. 9/11 Compensation: $500 to $8.6M. The Associated Press, Washington, D.C., June 15, 2004.

59. September 11th Compensation Fund Regulations Announced. Press Release. Department of Justice, December 20, 2001, www.usdoj.gov/opa/pr/2001/december/01_ag_658.htm.

60. Sebok, Anthony. The Final Rules for the September 11 Victims Compensation Fund: Are They a Laudable Model, Or A Large Mistake?, Findlaw's Legal Commentary, March 25, 2002, http://writ.findlaw.com/sebok/2002 0325.html.

61. Ramirez, Anthony. Joining Forces to Care for the Ill and Uninsured of 9/11. New York Times, February 25, 2007, p. YT 19.

62. Barrett, Devlin. 9/11 Compensation: $500 to $8.6M. The Associated Press, Washington, D.C., June 15, 2004.

63. Silverstein, Michael. Getting Home Safe and Sound. OSHA at Thirty-five. American Journal of Public Health, in press.

64. Greenhouse, Steven. Cleaning Companies in Accord With Striking Houston Janitors. New York Times, 11/21/06, p. A17.

65. Leary, Elly. Immokalee Workers Take Down Taco Bell. Monthly Review, Vol. 57, Number 5, October 2005, www.monthlyreview.org/1005leary.htm.

66. www.jwj.org/about.html and www.jwj.org/projects/wrb/history.html.

67. www.jwj.org/projects/wrb/history.html.

68. Unlikely pairing. Scientists, evangelicals share concerns on global warming. Associated Press, in Daily Hampshire Gazette, 1/20/07, p. B6.

69. Honoring Workers. AFL-CIO, www.aflcio.org/issues/safety/memorial/.

70. Faux, Jeff. *The Global Class War.* John Wiley & Sons, Inc. Hoboken, New Jersey. 2006, pp.157, 175.

71. ibid., p.161.

72. ibid, pp.175–177.

73. ibid., p.238.

74. *Death on the Job: The Toll of Neglect.* 13th Edition. AFL-CIO, Washington, D.C., April, 2004, p.17.

75. Jordan, Alfred D. II. A Study of the Profitability of Workers' Compensation Carriers in the State of Florida, 2001; and Death on the Job: The Toll of Neglect. 13th Edition. AFL-CIO, Washington, D.C., April, 2004, p.17, 18.

76. *Death on the Job: The Toll of Neglect.* 13th Edition. AFL-CIO, Washington, D.C., April, 2004, p. 18.

77. ibid.

78. Corporate Documents. Baron and Budd, P.C., www.baronandbudd.com/legal__services/asbestos/knowledge/documents.

79. Zerubavel, Eviatar. *The Elephant in the Room.* Oxford University Press, New York. 2006, p. 42,43, and Ending Legal Secrecy, New York Times, September 5, 2002, p.A22, cited in Zerubavel, Eviatar. *The Elephant in the Room.* Oxford University Press, New York. 2006, p. 42,43.

80. Leigh, J. Paul, Markowitz, Steven, Fahs, Marianne, Landrigan, Philip. *Costs of Occupational Injuries and Illnesses.* University of Michigan Press, Ann Arbor, MI, 2000, p. 12.

81. Schlosser, Eric. The Chain Never Stops. Mother Jones Magazine, July/August, 2001.

82. Fungicide Drift near Arvin. KBAK, Channel 29, cited in Take Action: Giumarra vineyard workers exposed to pesticide drift ... again, United Farm Workers, www.ufwaction.org, 9/6/2006.

83. Crowley, Dan. Worker error led to collapse. Daily Hampshire Gazette, 10/20/06, p. Al.

84. Barstow, David. When Workers Die. New York Times, December 21–23, 2003, p.l.

85. Leigh, J. Paul, Markowitz, Steven, Fahs, Marianne, Landrigan, Philip. *Costs of Occupational Injuries and Illnesses.* University of Michigan Press, Ann Arbor, MI, 2000, p.193.

86. Silverstein, Michael. Getting Home Safe and Sound? OSHA at Thirty-five. Chapter 2. Getting Home Safe and Sound—A Roadmap. Draft, p.4. http://defendingscience.org/newsroom/Safe-and-Sound.cfm.

About the Author

Patrice Woeppel, Ed.D. has had a long career in hospital, healthcare and mental health administration, as well as serving on many state and community boards, and as a hospital board officer. Dr. Woeppel has been involved in many civil rights and human rights issues over the years.

Index

978-0-595-48373-0
0-595-48373-9